Natural Alchemy:

Evolution of Life

Course 12-1

Natural Alchemy, Part 1

Evolution of Life

C. C. Zain

The CofL Press

www.light.org

THIS IS A BROTHERHOOD OF LIGHT BOOK
PUBLISHED BY THE COFL PRESS

Natural Alchemy: Evolution of Life is part one of the twelfth in a series of twenty-one courses comprising twenty-three volumes.

 Published in the United States by The Church of Light, Albuquerque, New Mexico.

www.light.org.

ISBN 978-0-87887-512-2

Foreword

In 1909, while attending a Brotherhood of Light meeting in Denver Colorado, Elbert Benjamine was approached by the Order to write The Brotherhood of Light lessons. In 1910 he agreed and began the process of writing under the pen name C. C. Zain. As each lesson was completed, it was published in serial lesson format and distributed to the members of The Brotherhood of Light. He completed writing all of the lessons in 1934. Eventually the serial lessons were formed into the books or courses we have today.

Until he passed away on November 18, 1951, the books were continuously edited to incorporate new information or to correct errors. After his passing to the inner plane, per his request, the books remained the same. Because the material contained in the Twenty-One Brotherhood of Light Lessons reflects the cultural milieu of the early twentieth century, there will undoubtedly be notions and ideas that have since been replaced with more highly evolved thinking. He predicted it would be so. For example, many of us today find we are put-off by sexist language that was commonplace during his era, when the generic use of 'man' and 'he' (and 'his', 'him', 'himself') was commonly considered gender-neutral. And while much of the science incorporated into the Lessons is still valid today, there are a number of scientific references that are no longer accepted. One example of this is the reference to the Piltdown man, which was found to be a forgery in 1952, the year after his passing. Zain also makes reference to the "ether" as a medium for the transmission of electromagnetic waves. While he points out that the concept of the ether was discarded by the scientists of his day as superfluous, he nevertheless portrays it as a useful concept. It is interesting to note that the concept has reentered scientific discussions in recent years.

Although there are a few things that seem archaic or incorrect, there is a wealth of useful concepts and information. For example, his theory of Astral Substance and Thought Elements or Thought Cells present a new paradigm for explaining how the unconscious mind works and how astrological energies work in stimulating particular thought elements. These theories shed new light on the Hermetic Axiom "as above, so below" and provide a unifying explanation for how the inner (formative) world interpenetrates and shapes the thoughts, feelings, emotions and events of life. These new theories are profound and worthy of exploration and deep consideration.

Undoubtedly we have much more to learn about astrology or the application of alchemy. Nevertheless, C. C. Zain's concepts of Directed Thinking and Induced Emotion were ground-breaking. Magic seems to be morphing into modern day psychology and the recognized power of visualization and positive thinking.

When it comes to the basic quest for "truth," none can compare. Zain advocates that we not accept something as true because he stated it in The Brotherhood of Light Lessons. Rather, he challenges each of us to verify the information for ourselves. It is this philosophy of continuing to ask questions that lead to further discovery. He often repeats in the Lessons: "the truth shall set you free." The quest for truth and a desire to liberate the human spirit keeps us moving forward in the journey of soul progression. As a philosophy of life, the integration of Hermetic tradition with modern science illuminates the path forward. The Brotherhood of Light Lessons shine as a bright beacon, elucidating the experiences of everyday life.

The Church of Light, 2013

Elbert Benjamine (C. C. Zain)
1882-1951
Author of the 210 Brotherhood of Light Lessons

C. C. Zain is the pen name used by Elbert Benjamine, the noted astrologer, naturalist and occultist, for those writings done under the aegis of The Brotherhood of Light during the years 1914-1951. This body of knowledge has become known as The Brotherhood of Light Lessons.

Elbert Benjamine was one of the most prolific astrological writers of the 20th century. In addition to the twenty-three volumes of the Brotherhood of Light series published under the penname C. C. Zain, Benjamine also wrote more than fifty books and hundreds of magazine articles. A scholar who mastered every physical science of his time, Benjamine brought the same vigor to his exploration of the metaphysical sciences. Like the writing of Bailey, Blavatsky, Steiner, and Heindl, the work of C. C. Zain (Elbert Benjamine) has impacted the lives of thousands of students of Western Occultism.

Elbert Benjamine was born Benjamin P. Williams in the small town of Adel, Iowa, December 12, 1882. He was a natural psychic and seer. As a youth his heightened awareness brought him into contact with those who had passed from this plane to the next. In the autumn of 1898, he began his esoteric studies. By 1900 he had contacted The Brotherhood of Light and began serious study of astrology. His father was a doctor and deacon in the Disciples of Christ church in Iowa, where the community strongly disapproved of any interest in astrology and the occult. For this reason, upon moving to Los Angeles, he changed his name to Elbert Benjamine in order to protect his family. In the spring of 1910 he gave his promise to write the Twenty-One Brotherhood of Light Courses covering astrology, alchemy and magic, under the penname of C. C. Zain. From that time until his death in 1951, he devoted his life's energy and personal resources to writing the lessons, and to establishing The Church of Light as a vehicle for disseminating The Brotherhood of Light teachings.

Preface

Religion should give instructions in optimum living. Optimum living embraces more than a few hilarious days, a few enjoyable weeks, or a few years of health and material prosperity which are followed by a long period of illness and misery. Optimum living gets the best out of life relative to its entire span. Considering the tremendous accumulation of scientific evidence that life persists after the dissolution of the physical, religion must embrace both life on earth and life beyond the tomb.

For living to best advantage after life on earth is done man must know as much as possible about the inner plane realm, about its energies and properties. And it is becoming increasingly evident that for him to live to best advantage while still in the physical form he must know as much as possible about these inner plane energies.

University scientists have demonstrated extrasensory perception. Man's soul often acquires information, usually unknown to himself, upon which he acts successfully to adapt himself to future conditions he could not have perceived through his reason and physical senses. This extrasensory perception, through which all information must be acquired after he loses his physical body, is equally valuable during and after physical life.

University scientists have also demonstrated psychokinesis. As man will have no physical muscles, and as objects of the inner plane do not respond to gravitation or physical pressure of any kind; after leaving the physical, to move or build anything, or to go anywhere, man must exercise psychokinesis. While still on earth he often is able to bring psychokinesis into play to heal the sick and amazingly demonstrate other desirable physical conditions. Because of this, the use of psychokinesis on earth is equally as valuable as its use after earthly life is done.

On the inner plane there is no air, no moisture, and no molecular vibrations which constitute heat. Thus after he leaves the physical he is not influenced by physical weather. He is influenced markedly by astrological vibrations, which constitute the inner plane weather. Though he may not be aware of it while on earth, the inner plane weather has as much or more influence over his life as the outer plane weather. Therefore, knowledge of how to forecast these astrological conditions and what precautionary actions should be taken relative to them, is equally important to man in the after earth life as it is while he still occupies a physical form.

Mankind is becoming too well educated to be guided either in religion or in its political views by blind belief in propaganda. More and more it is demanding demonstrated facts from those who advocate some economic or political system. And in due time it will demand demonstrated facts on which to base its religion. In 210 Brotherhood of Light Lessons the writer has striven to set forth as many such significant outer plane and inner plane facts, and the logical inferences to be derived from them, as possible.

The writer believes The Religion of the Stars will be the world religion of the future not merely from the facts and logical inferences presented in these 210 Lessons but because these facts will be supplemented by additional facts as fast as they are discovered and verified. The Religion of the Stars is not a static religion. It will progress as fast as there is progress in demonstrable knowledge.

This writer is not so foolish to believe that what has already been published in the 210 Brotherhood of Light Lessons is the last word, or that no errors have been made in them, or that new demonstrated facts may not make necessary some revision of the ideas there presented. He all too well remembers that when he went to college, the atom of each of the many chemical elements was indivisible, unchangeable and indestructible. Einstein had not yet published his Theory of Relativity. And four things which since his youth have so greatly changed civilization, as yet had no existence: automobiles, airplanes, cinema and radio.

While he is still on earth he will do all in his power to acquire new significant facts and revise the Brotherhood of Light Lessons to include them. When he has passed to the next plane undoubtedly new significant facts will be discovered that should be included in The Religion of the Stars. However, as orthodoxy will certainly try to get sufficient control to slant them into conformity with orthodox opinion, he believes the Brotherhood of Light Lessons as he leaves them should remain unchanged.

It would be unethical for someone to insert opinions or discoveries in these lessons and not take both the credit and the blame for them. The writer does not want the credit for the ideas or the errors of some other person. He asks that the printed pages of each lesson be left as he has last revised it.

However, in reprinting, it is easy to increase any lesson to 36 or 40 or any multiple of four pages. He suggests, therefore, that any errors he has made, or new discoveries, or logical opinions derived from these discoveries, be set forth and elaborated in an appendix following the 32 (original format) pages of the lesson which it is thought should be thus amended. Before this is done the writer of the appendix should submit what he has thus written to The Church of Light Board of Directors and secure their approval. And his name should appear in the appendix as the author of such commentary.

The author of the 210 Brotherhood of Light Lessons desires that they be permanently retained as the Stellarian Beliefs as he has written them up to the date of his physical demise, and that subsequent amendments should be credited to the persons who make them.

C. C. Zain
(Elbert Benjamine, 1882-1951)

Written in August, 1951
Los Angeles, California

About this Book

Natural Alchemy: Evolution of Life is part one of the twelfth in the twenty-one Brotherhood of Light course series by C. C. Zain on the Hermetic Sciences, Astrology, Alchemy, Tarot, Kabbala, and the Occult. The material in The Brotherhood of Light course books was written in the first half of the twentieth century. Chapters were first issued as individual serial lessons. Ultimately the serials were bound as chapters into twenty-three volumes and are known as the "Twenty-One Courses."

The student format, 8½ x 11 with wide margins, is designed to facilitate study. Study questions, located in an appendix, are designed to facilitate taking a final exam that Church of Light members are entitled to submit.

The birth charts in this book were selected by Elbert Benjamine (C. C. Zain) because they represent notable personalities of his day. Many of these will be unfamiliar to most people today. They are included for the astrological correlations to events mentioned in brief biographies that provide valuable examples.

Recognizing the importance of language, The Church of Light wishes to acknowledge that some of the Brotherhood of Light course books may contain language that is today considered sexist, racist or elitist. As with all great leaders and teachers Elbert Benjamine (pen name C. C. Zain) was a product of his time. The Church of Light is, however, bound to C. C. Zain's request that the material not be altered. Instead he suggests that any new facts or changes be introduced in the form of amendments. For his position on this matter, see the "Foreword" in this book.

For additional information about The Church of Light, the publisher of this material, please go to light.org.

Table of Contents

Illustrations

Key to Brotherhood of Light Serial Lessons[1]

3 BRANCHES OF STUDY

COURSE TITLE	COURSE NUMBER	SERIAL LESSON NUMBER
ASTROLOGY		
Astrological Signatures	2	1-5, 20, 21, 46, 47
Spiritual Astrology	7	71-83
Horary Astrology	8	36, 86-92
Natal Astrology (2 books)	10	19, 103-117
Mundane Astrology	13	141-150
Weather Predicting	15	190-196
Stellar Healing	16	197-208
ALCHEMY		
Spiritual Alchemy	3	49-54
Mental Alchemy	9	95-101
Natural Alchemy (2 books)	12	125-140
Occultism Applied	14	151-162
Cosmic Alchemy	17	164-172
Organic Alchemy	19	209-215
Personal Alchemy	21	216-225
MAGIC		
Laws of Occultism	1	39-45
Ancient Masonry	4	6-18
Esoteric Psychology	5	56-67
The Sacred Tarot	6	22-33, 48
Divination & Char. Reading	11	118-124
Imponderable Forces	18	183-189
The Next Life	20	173-182

[1] The Brotherhood o f Light Lessons were originally published as individual serial lessons and later incorporated into book format.

Chapter 1

Origin of the Earth

Serial Lesson Number 125
Original Copyright, 1949
Elbert Benjamine
a.k.a. C. C. Zain

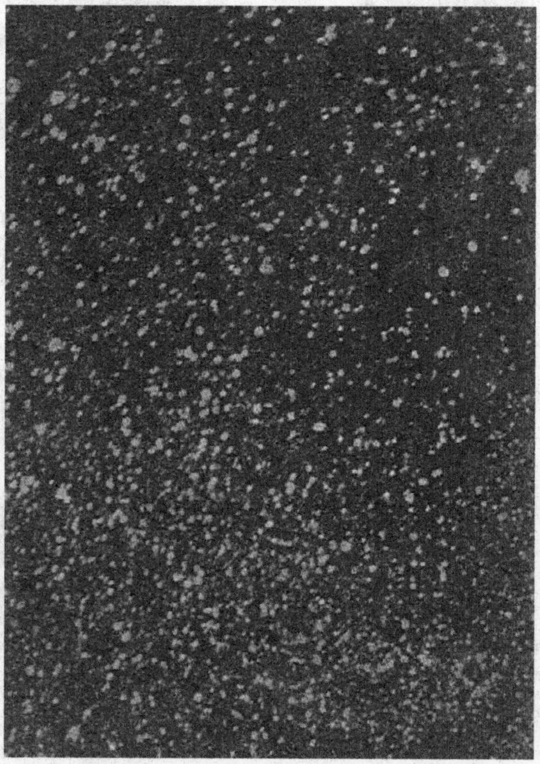

Apparently Empty Space

Photograph with high-powered telescope of apparently empty space.

Extra-Galactic Nebula - Front View

Front-view photograph with high-powered telescope of extra-galactic nebula.

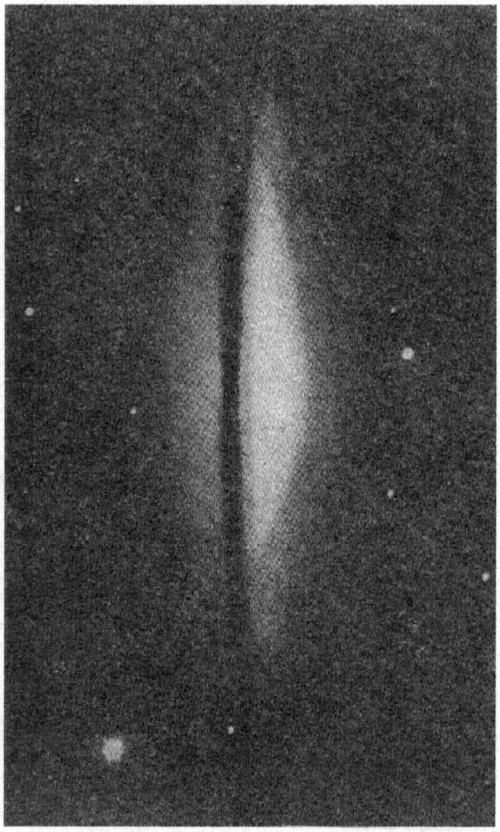

Extra-Galactic Nebula - Edge View

Edge-view photograph with high-powered telescope of extra-galactic nebula.

Chapter 1

Origin of the Earth

BECAUSE it regulates his conduct, it is difficult to conceive anything of greater importance to man than his religion. Even though it be unconscious and ill-defined, every person has a philosophy of life which determines his thoughts and actions in the face of innumerable circumstances by which, from day to day, he is confronted. He is aware of his own existence, and he is aware of the existence of other entities and forces than himself. It is impossible to ignore these other energies and persons and things, for continued existence demands they be recognized and due allowance made for their value as life-supporting or life-destroying factors. Thus each person is constantly called upon to make an adjustment to meet the requirements of contact with this or that entity, person or force. The nature of the adjustment so made depends upon his philosophy of life, conscious or unconscious, and which, although he may be inclined to reject the word may, in its broadest sense, be termed his religion.

Because his conduct is regulated by his philosophy of life, perfect conduct depends upon a perfect religion, and a perfect civilization, depending as it does upon man's conduct, is impossible without a perfect philosophy. Such a perfect religion must define man's proper relation to all other entities in the universe. But we can only formulate a system of relations between man and other men and entities from a knowledge of the man and his functions and a knowledge of other men and entities and their functions. Perfect relationship can only be determined from a complete knowledge of the man and a complete knowledge of other men and other entities. Such complete knowledge of man and such complete knowledge of all other entities and forces in the universe is not at hand. Consequently there can be no such thing as a complete and perfect religion.

Perfection is too much to expect, for this is a progressive existence, and progress demands new adjustments from time to time. We can conceive of no limit to the information it may be possible for man to obtain. New information must be assimilated by any adequate religion, and is thus being assimilated by The Religion of the Stars as fast as it is amply verified. To be adequate, religion must be based upon as complete a knowledge of man and other entities as possible. It must not remain stationary. It must progress even as knowledge progresses.

Such knowledge is not to be obtained by theorizing, nor by building fantasies—much as the human mind is prone to follow this line of minimum resistance—but by painstaking observation of man and other entities, by careful research into the life histories of men and all other living things, by investigating the psychology of the human mind and that of other creatures, and by extensive research on other than the physical plane. Only upon the most inclusive knowledge of both the inner plane and the outer plane can we hope, at last, to build a theory accurately and truthfully portraying, in so far as present day circumstances will permit, man's proper relation to all.

The theory of human life and conduct so derived, which we call The Religion of the Stars, must be logical inferences based upon as many inner-plane facts and outer-plane facts as it is possible to acquire.

Nature, in her various ramifications is so vast that any one man may personally investigate thoroughly only a small fraction of the whole. Hence, men of science usually specialize and devote their chief endeavor to some single section, or small department, of nature. They attain great skill in research, and gain much precise knowledge concerning the minute region of the universe to which they direct lifelong attention. But because of this they are hardly better fitted to pass judgment on the whole than the man who has made no study. To pass competent judgment about the universe, there must be available for comparison facts, not merely about a single department, but fundamental facts regarding as many departments and sections of nature as possible.

Nor are scientists who specialize in one field competent to pass judgment on the facts in an unrelated field of science. Yet often they are prone to do so. Comments Jan Ehrenwald, M.D., in an article on the Neurobiological Aspects of Telepathy, appearing in the October 1948, The Journal of the American Society for Psychical Research:

"An increasing amount of experimental evidence of telepathy and related phenomena has been accumulated during the past fifteen years. The problem no longer is to convince the skeptics who, by their stubborn disbelief, prove only their inability to digest food for thinking that had not been included in their diet during their childhood years."

During the first thirteen of the years mentioned, up to 1947, Duke University Laboratory alone conducted over one-million trials of extrasensory perception; other university laboratories, following similar methods reported over two million trials, and there were something over a million trials, with responses from over 46,000 subjects made by the Zenith radio program in the winter of 1937-38.

Dr. J. B. Rhine, of Duke University, the instigator of the experiments there conducted, at the commencement of his book, The Reach of the Mind (1947) says: "Henceforth I will assume that science will in time accept ESP and PK and that Psi is a normal human capacity, nonphysical in nature."

No intelligent person who follows the scientific experiments with psi phenomena over the years can fail to be convinced that the soul continues to live beyond the tomb, and that its personality there is in essential respects that which it exhibited while on earth. Nor can any intelligent person who learns to erect and progress a birth chart fail to be convinced that the inner-plane weather, which consists of astrological energies, has as much or more influence over human life than the physical weather.

Man lives in, and is influenced by, both an outer-plane world and an inner-plane world. And there has been a vast amount of observation, carefully checked, which indicates that if we consider man to consist of his physical body, his astral body, his mind or soul, and the thoughts which he thinks, the inner-plane environment—which includes objects, the actions and thoughts of intelligent entities, and astrological energies—has as much influence over his thoughts, feelings and behavior as do all outer-plane conditions and energies, including the influence of his associates.

The facts thus far discovered relative to the inner plane, its energies and the way its inhabitants live are set forth in the other 20 Brotherhood of Light courses. In this course a survey of the fundamental facts having to do with the progress and development of life and religion on earth will be made, in so far as it is possible to cover so vast a field in so limited a space. Necessarily this outline must be brief, and because the scope of nature is so limitless, that which is included is as a pebble to a mountain to that which is both interesting and important that must be omitted. Yet if I can sketch even so brief, though clear a picture of the processes that, according to the latest findings of science, have brought the world from its primitive stardust to where it is inhabited by civilized man, and show the steps by which his religion has reached the present stage, I shall be quite satisfied.

But in addition to explaining the processes by which life on earth has made constant progress, I believe I should also point out the factor which during 1900 years most retarded the acquisition of knowledge of the physical world, and which at the present time is the greatest hindrance to the dissemination of information about the inner world.

The Iron Curtain of Orthodoxy

Orthodoxy for centuries kept the Western World behind an iron curtain by means of the Inquisition. It no longer resorts to ruthless violence, but it persists in building an iron curtain that effectively prevents most from learning the facts. In their emotionally impressionable childhood people are taught they should make no investigation of religion. They are impressed that instead they should place reliance on blind belief. And to keep them behind this iron curtain of ignorance they are threatened with eternal suffering after earthly life is done.

In 1163, acting in conjunction with the Council of Tours, Pope Alexander III forbade ecclesiastics to study "physics or the laws of the world." Roger Bacon, over a hundred years later, failed to obey this order. He explained the rainbow and other natural phenomena by scientific methods. In 1278 the Franciscan order condemned his teachings, and later the Pope threw him in prison for 14 years. He was not released until he was 80 years of age.

In 1243 the Dominicans forbade every member of their order to study medicine or natural philosophy, and in 1287 forbade them to study chemistry. In 1380, Charles V of France, at the insistence of religious authorities, forbade any person to possess furnaces and apparatus necessary for chemical processes, and in 1404 Henry IV of England made a similar law, as did Venice in 1418.

Eufame Macalyane, a lady of high position, in 1591 was charged with "seeking the aid of Agnes Sampson for the relief of pain at the time of the birth of her two sons," and as church authorities held that woman should suffer pain in childbirth, she "was burned alive on the Castle Hill of Edinburgh."

Later, when James Young Simpson advocated the use of anesthetics in childbirth he was furiously denounced by church authorities and told that what he proposed was "to avoid one part of the primeval curse on woman."

In England, Scotland and New England as late as the eighteenth century it was considered "flying in the fact of providence" to attempt to prevent smallpox. God had sent it in his judgment, and to attempt "to avert it is to provoke Him more."

The introduction of quinine in Europe was bitterly fought by orthodox Protestants, and it was not used in England until 1653. The natives of South America held coca in great esteem. But the Second Council of Lima, in 1567, condemned its use, and in 1569 a royal decree was issued declaring that "the notions entertained by the natives regarding it are an illusion of the devil."

In spite of the fast accumulating evidence that the earth is a sphere, Luther, Calvin, Zwingli and Melanchton, among the Protestants, held steadfastly to the Bible dictum that its shape is four-cornered. As to the possibility of people living at the antipodes, the early church held that to be heresy. Said St. Augustine, such teaching was to "give the lie to King David and to St. Paul, and therefore to the Holy Ghost."

Having set the date of creation, orthodoxy did all in its power to discredit the records of the rocks and the evidence of early man. Fossils to it were the remains of the flood of Noah, and the artifacts of primitive man were those of people living later than 4,000 B.C. There was a great outcry when in 1844 Robert Chambers published his Vestiges of Creation, still greater anger when on July 1, 1858, the papers of Charles Darwin and Alfred Russel Wallace were read before the Linnean Society of London, and a tirade of abuse in 1859 when Darwin's Origin of Species was first published.

It is not surprising that man in the past had erroneous ideas about himself, about the earth, and about life after physical dissolution. Nor at the present day do we possess all-important information on these subjects. It is not to criticize the ignorance of the past, or that of the present, that these matters are being mentioned. It is to point out how the iron curtain of fear erected by orthodoxy has impeded investigation and has made people afraid to recognize facts which are essential to their present and future well being. And to emphasize that this iron curtain of fear erected by orthodoxy must be lifted if man is to live to best advantage.

Raising the iron curtain of orthodoxy has been difficult even where the physical world is concerned. But to raise it where inner-plane facts are concerned, even though this is equally important, is a far harder job.

Relative to the earth itself, Archbishop Ussher of Ireland in 1654 declared from his study of the Scriptures that Creation had taken place in the year 4004 B.C. This was then inserted in the authorized version of the Bible. It became, along with the notions of other church men of previous times, part of the orthodox dogma.

As such it discouraged all research as to the actual age of the earth, and brought denunciation on the head of any person who made such investigation. Yet in 1778 Buffon had the courage to estimate the age of geologic time as 75,000 years. In 1860 Phillips shocked his contemporaries by placing the age of the earth at 60 million years. But by 1900, as the result of studies of time taken by erosion to level mountain chains, build sedimentary rocks, and to cut canyons thousands of feet deep in these rocks, geologists usually accepted 100 million years as the probable age of the earth.

They found that in recent years Niagara Falls had retreated five feet a year, wind erosion at certain places in the Gobi Desert was at the rate of five feet in 156 years, the bluffs of till along the shore of Cape Cod were receding at the rate of from one to three feet a year, peat was accumulating in marches at Lynn, Massachusetts, at the rate of three to four inches in fifty years, the rate of down cutting of the Nile at the Cataracts has been at the rate of 25 feet in 4,500 years, and certain glaciers deepen their cirques 0.57 millimeters a year.

Where periods are not too long, estimates derived from such rates of change are fairly accurate. For shorter periods where each annual layer of sediment remains distinct enough to be counted, as are rings of a cut tree, there is a still more precise measure. For instance, such annual varves of clay representing 20,000 years or more have been counted in many lake beds.

There was no precise measure of the longer periods of time which had elapsed since the rocks of earth were formed until after 1902. That year radium was discovered, initiating the study of simple elements which are subject to spontaneous disintegration. About twenty such atoms are known in nature which expel parts of their nuclei. The process is quite regular, very slow, and the speed of such disintegration cannot be changed by heat, pressure, or any other known external influence. In one year heavy uranium spontaneously loses one atom out of 6,570 million; light uranium (actino-uranium) loses one atom out of 1,030 million, and thorium loses one atom out of 20,000 million.

The atoms thus lost go through various changes some of which require only a fraction of a second and some of which not less than a million years. These changes follow a regular known sequence, and finally become stable nuclei each of which is an isotope of lead. Lead of non radioactive origin always contains a slight proportion of an isotope with an atomic weight of 204. This never appears in the course of radioactive disintegration. Instead, the isotopes of lead resulting from the disintegration of heavy uranium, light uranium and thorium have atomic weights of 206, 207, and 208. Therefore, natural lead in rocks can readily be distinguished from lead due to radioactive disintegration.

If a rock containing ordinary uranium was formed a billion years ago, about 14 percent of the original uranium atoms will have disintegrated and been replaced by an equal number of atoms of lead. The weight of the lead will be 12 percent of the original uranium, and the weight of the helium liberated in the disintegration will be two percent. The ratio between the amount of lead and the amount of uranium found in the rock provides a faithful and very sensitive clock for timing the age of the rock.

On the basis of such calculations, uranite deposits at Great Bear Lake are 1,375 million years old, and dykes in the Black Hills at Keystone, So. Dakota, are 1,420 million years old. The oldest dated rocks thus far found are a formation in Carelia, Russia. They are 1,800 million years old. These dated rocks, however, are not the oldest, for they consist of intrusive material thrust through older surrounding rocks which contain no radioactive minerals. It is assumed, therefore, that the oldest rocks are about two billion (2,000,000,000) years old.

As to how old the earth was before it was cool enough for its crust to solidify as rock, science is today in as much of a predicament as it was at the turn of the century in estimating the age of rocks. To replace the nineteenth century Laplacian nebular hypothesis, there has appeared in the present century the tidal friction theory of Jeans and Jefferies, the planetesimal theory of Chamberlin and Moulton, and a still newer theory of Jefferies. The tidal friction theory and the planetesimal theory assume the close approach of two stars, and the newer theory of Jefferies a side swiping collision between them.

In each case one of the stars, or what is left of it after the side swiping collision, passes off in a hyperbolic orbit. The close approach or the collision is supposed to have caused the ejection of material from the star which we call the sun. The planetesimal theory postulates that the ejected material condensed and solidified quickly into small objects known as planetesimals, and that the planets were formed by the gathering together of the small planetesimals about nuclei. The tidal theory and the later one of Jefferies assume the planets were formed by the condensation of large masses of hot diffused material.

On the basis of these theories it has been assumed the earth was in existence about a billion years before it solidified. Thus a geological time clock in the American Museum of Natural History made in 1931 gives the gaseous stage of the earth as 400 million years, at which time, 2,600 million years ago, meteors began to fall. Then followed 600 million years of Azoic time before the earth's crust formed about two billion years ago.

However, not only have discrepancies been found in each of these three theories, but the most eminent astronomers in this year of 1949 believe our universe cannot be much more than two billion years old. Even as in the opinion of science since the turn of the century the age of the earth has increased, so in its opinion has the age of the universe decreased. Since the commencement of this century Milne demonstrated that the sun cannot have existed more than 5,000 billion years, and Eddington held that it seemed unlikely the age of the stars, systems and galaxies could go back more than 10,000 billion years.

The spiral nebulae, which are galaxies beyond our own galaxy of the Milky Way, are each made up of hundreds of millions of stars. And they all appear to be moving away from us and from each other at speeds that increase in proportion to their distance. If this expansion has been continuous at the indicated rate, not much over two billion years ago all of the observable universe must have been concentrated near one point. The theory is that the universe in its present form resulted from atomic explosion within this once compact mass.

The movements of stellar bodies can be observed only through the aid of light. The red shift in the spectrum indicates the movement of these galaxies away from us. One of the problems astronomers hope eventually to solve by means of the 200-inch telescope on Palomar Mountain is whether the red shift is due to some hitherto undiscovered law of nature, or actually indicates the universe is expanding.

In 1940 work was begun on the 4,000-ton atom smashing apparatus at the University of California. Its construction, interrupted by World War II, was resumed in 1945. Then the original design had to be modified to adapt it to the newly-recognized principle of frequency modulation. It had to be modified to make correction for the effect indicated by the Special Theory of Relativity which makes particles increase rapidly in mass as their velocity approaches that of light. And it is possible—as relativity affects not only gravitation, as is also demonstrated by psychokinesis, but distance, as demonstrated by its apparent lack of interference with extrasensory perception and telepathy, and time, as demonstrated by the ability of extrasensory perception to perceive events both in the past and the future—that light rays traveling through space for hundreds of millions of years are subject to laws not yet recognized.

Not only is matter converted into atomic energy for many present day uses, but it has been proved that when an electron meets a positron both are annihilated with the release of gamma radiation. And it has been proved that the action of a gamma ray on the electromagnetic field can call an electron and a positron into existence. That is, matter can be created from energy. Although other scientists do not all agree with him, Professor R. A. Millikan, one of the two outstanding authorities on the cosmic ray, holds that in their primary form they are a result of the synthesis of certain types of atom in outer space and that they represent the "mass defect" energy of these atoms.

At least we now know, due to progress in photography, that interstellar space, two decades ago believed to be empty, contains a tremendous amount of microscopic matter. Jan Oort, in 1948 president of the International Astronomical Union, calculates that the interstellar gas and dust contains as much matter as all the stars. In the Milky Way, only a minute segment of the observed universe, this comes to 300 million million times the mass of the earth. Hydrogen and helium are prevalent, and some think these two gases are the building blocks from which by nuclear synthesis many of the other elements were made.

The most plausible theory of today (1949) is that our solar system once consisted of a vast dust cloud which began to condense under gravity. As the Milky Way itself rotates, dust clouds within this area would also rotate. According to the law of conservation of angular momentum, as it shrunk in size it would rotate more rapidly. With resistance diminishing and gravity increasing the cloud would collapse faster and faster, its final collapse to a size equal to that of the solar system requiring only a few hundred years. Due to increased pressure the temperature would rise tremendously, and in its final white-hot phase the sun would become hot enough to start the nuclear reactions among carbon, hydrogen and helium which some believe is the source of its continued heat.

It is held that some of the revolving cloud condensed into minor clouds and in the final collapse of the sun these were left stranded to become planets. A number of known laws, including that of the pressure of light from the sun, are called in to account for their behavior and that of their satellites. According to this theory, neither the earth nor the universe need to have been in existence over about two billion years, and thus its materials could have resulted from an atomic explosion of a central mass not much longer than that ago.

The Part Played by Psychokinesis

The evolution of life, as will be made plain in subsequent chapters, is from the simple toward the more complex. But inorganic evolution, in obeyance to the Carnot-Clausius law, moves successively from the more complex to the simple. According to this law every successive inorganic state entails a definitive decrease in its available energy. Unless it is able to borrow energy from some other system—such as the astral plane—the material universe is running down and in time will reach a state of inert symmetry.

From whence came these molecules, atoms, protons, neutrons, positrons, electrons, mesons, photons and possibly neutrinos of the inorganic world which are now running down? And from whence came life which is now evolving from the simple to the complex? Lecomte du Nouy in 1947 showed by probability calculations the inconsistency of believing the appearance of the first living cell to be due to a chance combination of inorganic molecules.

Psychical researchers have found that whenever physical conditions were present that would permit the manifestation of intelligence, that intelligence always was present there to manifest itself. In other words, there is an inner plane, nonphysical in nature, and not subject to physical laws, where intelligence of incalculable grades at all time persists. It is on this plane that the unconscious mind or soul of man exists while it functions through his physical body, and it is on this plane that it will continue to function and develop after the dissolution of the physical vehicle.

There is no evidence of the existence of a God who has human frailties, a God who unjustly can be persuaded to favor some special nation, Who punishes those who strictly follow religious doctrines in which they believe, but which do not conform to those of some other creed, and Who can be cajoled or bribed into granting special, and quite unjust favors to those who give Him praise and beseech Him in prayer.

But that there is an all-pervading Super-Intelligence which can be contacted there is much ESP evidence to prove. But not a God of whim and cruelty, such as would condemn to eternal suffering those who had no opportunity to embrace, or whose early conditioning in belief determined by environment over which they had no control caused them to reject, some special faith.

If, as has been amply demonstrated in university laboratories, the mind can manipulate physical objects, and if, as psychical researchers have demonstrated, mind on the inner plane with various grades of intelligence is at all times present, there seems good reason to conclude that the trend of evolution is in response to the psychokinetic power of images within the mind of the all-pervading inner-plane Super-Intelligence.

Under the jurisdiction of the overall Super-Intelligence, lesser intelligences on the inner plane perform their psychokinetic functions. According to the old Hermetic teaching, substance, motion and intelligence are inseparable and eternal. According to this teaching the positron and the electron which can be manufactured from the energy of an electromagnetic field are merely the high-velocity astral substance of the inner plane which has been slowed down through polarity.

Lecomte du Nouy in his book Human Destiny, has pointed to the fact that material science has utterly failed to account for life even in its most elementary form. To account for it, and to account for its evolution, the existence of an inner plane where mind resides and exerts a formative pressure on the lower-velocity physical substance must be recognized.

The Material Universe

After studies with the 200-inch telescope on Palomar Mountain, which start this year, have been completed, there will be much revision of present views about the universe. This glass will reach twice as far into space as the 100-inch telescope on Mt. Wilson, which is the largest used previous to this year.

According to calculations made from observation with it and smaller telescopes our sidereal universe, or galactic system embraced within the Milky Way, is in the shape of a lens between 200,000 and 300,000 light years in maximum diameter and about one-tenth of this in thickness. It is estimated to contain about 100 billion stars.

Within this main galactic system in the vicinity of the sun is a local system which also is somewhat lens shaped, having a maximum diameter of about 20,000 light years. The sun is about 275 light years from the center of this local system, which inclines to the plane of the main system at an angle of about 12 degrees. The local galaxy is about 65,000 light years from the center of the main galactic system. In this main galactic system there appears to be a central condensation of material in the general direction of the constellation Sagittarius.

Beyond our sidereal universe are between 50 million and 100 million extra-galactic nebulae within the one-billion light-year observational scope of the 100-inch telescope. These are of three types: spirals, elliptical and irregular. About 77% are spirals, 20% elliptical and 3% irregular. It has been proved that these extra-galactic nebulae are immense groups of stars. It is believed by many that viewed from the outside our sidereal universe, or Milky Way, which appears to be rotating, would appear as a huge spiral nebula, such as the spiral nebula in Andromeda appears to us. If this is the case, although other spiral nebulae are tremendously large, that of Andromeda, measuring 28,000 light years (light travels 186,284 miles per second according to measurements in 1942) in greatest length, our Milky Way is the largest of any of the galactic systems thus far measured.

The Stars: The stars which are found both in the extra-galactic systems and in our own galactic system, are undergoing a process of evolution. Further advances in nuclear physics will shed light on the changes they undergo. In their early stage they are supposed to be of the type known as Red Giants, and in their last visible stage they are supposed to be of the type known as Red Dwarfs which, while so much smaller, have approximately the same mass as the red giants. The largest red giant thus far measured is the star in the constellation Scorpio known as Antares. Measured with the stellar interferometer it shows a diameter of 390 million miles. As the earth is only 92,897,000 miles mean distance from the sun, and Mars is only 141,500,000 miles mean distance from the sun, if Antares were to occupy the place of the sun, both earth and Mars would be far interior to its surface.

The giants are largely composed of diffused gas at low temperature. As they contract they are supposed to get hotter and hotter until they reach the blue-white stage, when they start cooling off. Like an iron in a furnace they first become red, then orange, then yellow, then yellow white, then white and finally blue white. To conform to the two billion year expanding universe theory they reach the blue-white stage quickly, from then on cooling gradually. Thus the red giant Antares, 480 times the diameter of the sun, has a density of only 0.000,000,3, and a temperature of only 3,100 degrees. But the orange giant Capella in the constellation Auriga, with a diameter only 12 times that of the sun, has a density of 0.002 and a temperature of 5,500 degrees, and the blue-white B Centauri, with a diameter 11 times that of the sun, seems to have reached the maximum temperature before it starts to cool. Its density is 0.02 and its temperature is 21,000 degrees.

After a star reaches the blue-white stage witnessed in B Centauri it begins to radiate heat faster than its continued condensation generates it. From then on radiation pressure acts against gravitational contraction and tends to hold the mass in approximate equilibrium and permits it to cool slowly. Growing smaller and smaller it passes in reverse order through the colors white, yellow white, yellow, orange and red. Finally it becomes black and invisible. There are supposed to be many more black stars in our universe than those that emit light.

Sirius, in the constellation Canis Major, with a diameter 1.8 times that of the sun, has cooled to the white stage with a density of 0.4 and a temperature of

11,200 degrees. Our sun, an orange dwarf star similar in color to the orange giant Capella, has a temperature of 6,000 degrees, only 500 degrees higher than Capella, but has a density of 1.4, which is tremendously greater.

Antares has a luminosity 3,500 times that of the sun, Capella 150 times that of the sun, Sirius 26 times that of the sun, and Krueger 60A has a luminosity only 0.002 that of the sun. Our sun is well past its prime, for it is an orange dwarf. But it has a long way to go before it becomes one of the countless dead suns that clutter up the universe; even a long way to go before reaching the red dwarf stage of Kreuger 60A which, with a diameter of 0.3 that of the sun has a temperature of only 3,300 degrees, but a density of nine.

White Dwarfs: More than 99% of the observed stars fit into the evolutionary sequence of stars just set forth. But there are a few that do not fit into this general scheme. Because they are so small and faint they are difficult to locate. With the progress of nuclear physics more about their origin will be learned. The present theory is that they consist of atomic nuclei stripped of all external electrons and tightly packed together by gravitational compression. The first of these white dwarfs to be discovered was the companion star to Sirius. Its mass is about that of the sun, but although hotter than the sun, its luminosity is only about 1/360 as great. This means that about the same mass of material found in the sun with a mean diameter of 864,000 miles is compressed into the 30,000-mile diameter of the white dwarf. Its density, 30,000 times that of water compared with the earth's 5.58 times that of water, is such that a cubic foot of its material weighs 935 tons.

Multiple Stars: Not all the stars are single as is our sun. In many instances two stars form a system and revolve in elliptical orbits around their own common center of gravity. Sirius, the brightest star in the sky, is such a binary, its companion being the white dwarf just mentioned. Then there are triple stars, in which three stars form a system; and Theta Orionis, the star in the nebula in the sword of Orion, is composed of six stars. One multiple, as they are called, has been discovered that is really composed of a closely related system of 16 different stars. Such systems of more than one star often have one or more members of the family that are dark stars whose presence may be known only by their passing between us and some of the others.

Variable Stars: Any star whose light is known to fluctuate is called a variable star. At the present time 10,000 such variables are known, and it is estimated that not less than 5% of all stars are at least somewhat variable. One type, known as the Algol type because the star Algol behaves in this manner, shows a rapid diminution of its light at regular intervals. It is believed such stars have a dark companion, or one of less luminosity—and this has been definitely proved in the case of Algol—and as the two revolve the darker of the two periodically eclipses the brighter.

But there is another type of variable star whose irregular fluctuations cannot be explained in this manner. Some think they are dying suns, and that as a crust of thick vapor forms on the surface their light is shut off. Then, at irregular intervals, the molten interior bursts through and they again appear very brilliant. Others believe that they are suns traveling in a region rich in dark nebulae, and that when they enter successive clouds of dust or strike swarms of meteors the friction of the impact is the source of the added illumination.

Novae: Not only are there supposed to be many more dead and dark suns—suns that have cooled below the light-emitting stage—wandering about the heavens than there are luminous ones, but many new suns apparently are observed to be born. These are called novae. Many are detected each year, and it is estimated that ten or more reach a brightness of the ninth stellar magnitude or more annually. Photographic records indicate, however, that novae are not actually new stars, but are faint stars which for some reason suddenly increase in intensity. It is not uncommon for them to gain an increase of ten magnitudes, which means an increase of light intensity of 10,000 times.

At present no complete explanation is available, but with further advances in nuclear physics such an explanation may be forthcoming. The most commonly accepted present theory is that some faint star explodes and blows off its outer shell of gaseous material. This may be due to the release of atomic energy. As the shell would rapidly expand after leaving the star the displacement of spectral lines would be toward the violet, which is observed. Nor are these novae, whose brightness lasts only a short time and then rapidly fades confined to our sidereal universe, for they have also been observed in some of the extra-galactic systems such as the great spiral of Andromeda.

Star Clusters: Stars which move through space together in parallel paths are known as moving clusters. Many such moving clusters are known. The sun apparently is not a member of such a moving cluster, but at present is within the Ursa Major cluster, which has a diameter of 500 light years, with its members moving past us on both sides.

Out on the far fringe of our galaxy are the globular star clusters. Sixty-nine of these are known, each consisting of an immense number of suns closely grouped, comparatively speaking, in a globular system of stars. Shapley, of Harvard Observatory, finds that these clusters themselves, taken as a whole, form a huge flattened cluster, probably 250,000 light years in diameter, with its center about 75,000 light years from the sun in the direction of the constellation Sagittarius. Sagittarius seems to be the bullseye center of both the globular clusters and of our sidereal universe.

Magellanic Clouds: Also on the fringe of our galactic system toward the south pole of the celestial sphere appear two luminous roughly circular objects. They are called Magellanic Clouds. Occupying areas singularly devoid of bright stars, the larger covers a space of 42 square degrees and the smaller covers 10 square degrees. They appear as a promiscuous intermingling of star clouds, star clusters, and gaseous nebulae. Little is known at present concerning them.

Galactic Nebulae: The nebulae within our sidereal universe, or Milky Way, are of three different types: dark nebulae, diffuse nebulae, and planetary nebulae. Several hundred dark nebulae composed of immense dust clouds and possibly other dark material have been photographed. They range in size from small spots up to the great black rift which can be seen with the naked eye that extends almost a third of the way around the Milky Way from Cygnus to Centaurus.

The diffuse nebulae have no more specific shape than the dark nebulae and appear much like clouds of vapor. Their density is very low, and apparently they differ from the dark nebulae only in being luminous through usually being associated with a star which illuminates them.

Planetary nebulae are so called because they have much the same appearance in a telescope as a planet. In most instances they have a star of the hot blue spectral class at their center which is associated with their radiation. These nebulae are rotating and are probably composed of ellipsoidal masses of gas.

Asteroids: According to Bode's Law there should be a planet between Mars and Jupiter in our solar system. But instead, in this orbit travel swarms of smaller bodies, perhaps 150 having a diameter greater than 50 miles and the majority having a diameter between 50 and 20 miles or less. The most widely accepted theory is that there was once a planet there, but that it broke up due to atomic explosion or to collision with another heavenly body.

Comets: Comets are the largest members of our solar system, but their mass is very small. A definite connection has been established between comets and meteors. A comet was observed to disintegrate and a meteor shower took its place. Often as a comet approaches the sun the pressure of the light from the sun causes fine vapor-like material to stream out in a direction away from the sun. The earth passed through the tail of a comet twice during the last century and nothing was felt. The head of a comet, however, consisting of a swarm of iron meteors is more formidable. Not long ago such a meteor flattened a big forest in northern Siberia, and Meteor Crater, near Canyon Diablo, in northern Arizona is supposed to have been formed by a monster siderite meteor of some 500 feet in diameter, that was one of a flock that formed the nucleus of a large comet that struck the earth not more than 5,000 years ago.

Meteors: In addition to the wholly metallic siderites, there are sideriolites composed of both metallic and silicate materials, and aerolites composed almost entirely of silicates. Recent calculations and observations indicate that about half the meteors are visitors from interstellar space, having their origin outside our solar system.

Zodiacal Light: This light is seen in the west after evening twilight in the spring and in the east before morning twilight in the autumn. It is a pearly radiance that, cone shaped, slants up from the sunset glory, or from before sunrise, sometimes almost to the meridian, but always following the zodiac, or path of the sun. In the northern hemisphere, therefore, it leans to the south. It is due to sunlight reflected from cosmic dust.

Aurora Borealis: This phenomena commonly seen in high latitudes is an electrical discharge in the ionized air exhibiting the characteristic spectrum lines of the rarer atmospheric gases. Sunspot maxima, with their eleven-year cycle, are always accompanied by maxima brightness and frequency of these Northern Lights which are always coincident with disturbances of terrestrial magnetism. At other times they are most commonly seen in March and October.

The Moon: Some scientists believe the Moon was formed at the same time and in the same manner as the earth. Others hold that after its formation tidal strain caused the earth to bulge until the Moon broke off, and the back reaction of the lunar tides caused the Moon gradually to recede, and the solar and lunar tides, acting like a brake shoe against the earth's rotation, caused the earth to slow down. The rate of slowing down, according to the records of eclipses preserved by the ancients, indicates that at present the day is lengthening about 1/1000 second per century, or a minute in six million years. This might mean the Moon broke off 10 billion years ago, or if all possible allowances be made for greater tidal retardation when the Moon was closer, the date cannot be less than one billion years ago.

Chapter 2

Origin and Development of Plants

Serial Lesson Number 126
Original Copyright, 1949
Elbert Benjamine
a.k.a. C. C. Zain

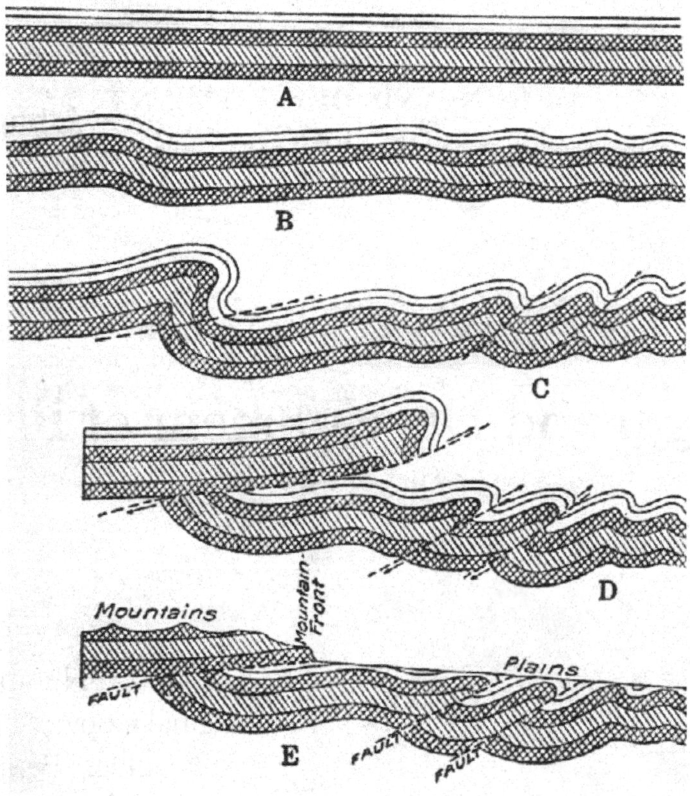

Glacier National Park Rock Formations

Diagram by U. S. Department of Interior showing, "How internal pressure transformed level rock into the tumbled masses of Glacier National Park. "A" represents the original water-laid rocks; "B" the first yielding to internal pressure; "C" the great folds before the break came; "D" and "E" the way the western edges overlapped the eastern edges when the movement ceases."

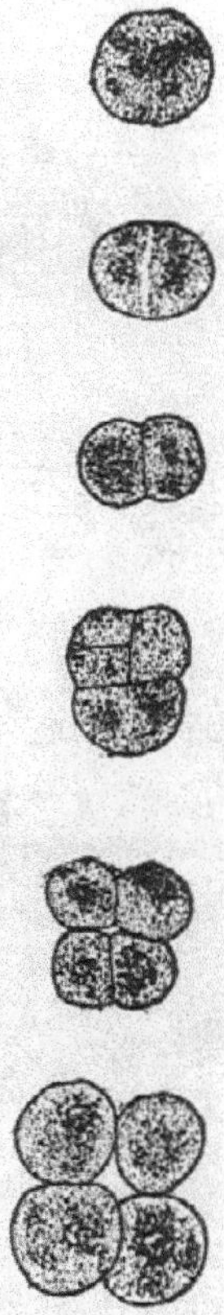

Cell Division

Diagram showing multiplication by
cell-division of single-celled green algae.

Elbert Benjamine (C. C. Zain)

Snapshot of Elbert Benjamine (C. C. Zain) taken in July, 1924, as he viewed the erosion in Bryce Canyon, Southern Utah (now a National Park). The larger tree at his right is a lodgepole pine.

Chapter 2

Origin and Development of Plants

IT is believed that after the moon broke away from the still hot earth it had a diameter of about 8,100 miles. Since that time it has shrunk in size, partly due to loss of heat, but to a still greater extent due to an internal rearrangement of its molecular structure, until its present diameter is but 7,918 miles. During the cooling the viscous material slowly continued to boil, and as a crust would form, hotter material from the interior would break through and the heavier portions of the crust would sink toward the center.

The fact that solid rock does not float on molten rock like ice on water, but tends to sink, is but one argument against the old notion that there is a crust at the present day supported by a molten interior. Undoubtedly the heavier materials would sink to the center of the earth. It has even been suggested that, as gold is one of the heaviest metals, the earth has a core of almost solid gold. This, however, is mere speculation. But from what is known of the weight and density of the earth, and of other celestial bodies like meteors, it is not improbable that the earth has a core some 4,000 miles in diameter mostly composed of iron. Other metalliferous and basic rocks, due to their relative weight, would lie above this iron core, while the acid rocks, chiefly granites, being much lighter, would rise to the top of the molten metal.

It is now quite well established that the continents are built of much lighter material than the ocean beds, the ocean bottom being of basaltic rock some 3% heavier than the granitic rocks forming the continents. Some now think that the region of the Pacific Ocean is where the moon broke off from the earth, the opposite section of the earth being where the tidal bulge was simultaneously formed on the other side of the earth. This then became the region where the granite frosting floating on the plastic and heavier basalt beneath, like the frosting on a custard pie, was left after the moon broke out from close to the surface on the other side.

Professor Wegener advances a theory that is finding considerable acceptance, that the continents were anciently much nearer together than at present. Nearly all geologists now believe that underlying the crust of the earth, say some 60 miles below sea level, there exists between the more metallic interior and the outer cover a rather thin layer of basaltic rock. It is also known, because the process has been duplicated in the laboratory, that under the pressure and heat known to exist in the flowage zone—below about 60 miles depth—that the rock flows like ice in a glacier, through recrystallization. While still as rigid as steel, it nevertheless flows under pressure without breaking, just as a piece of hard pitch or asphalt may be made to assume any form without breaking by subjecting it to gradual pressure. Yet the same pitch or asphalt or the same rock, if subjected to quick strain, will break like glass.

Professor Wegener believes that South America, Antarctica, Australia and India were once much closer to South Africa than at present. In fact, they seem to fit to South Africa when their shore lines are brought together. And in like manner North America seems to fit Europe. This proximity of the Old World and the New does not fail to take into account Atlantis, for its existence is well established. Certain it is, from the similarity of their flora and fauna, that the continents were connected by land at no very distant date, geologically speaking. Professor Wegener believes that these blocks of granite frosting, floating on the viscous basalt beneath, broke apart in the Tertiary Epoch, and that America drifted westward away from the Old World. Such westward floating might naturally arise from the eastward rotation of the earth on its axis, and from other known forces.

Greenland at the present time is moving away from Europe at the rate of 50 feet a year. The American continents in their westward movement are supposed to have buckled up the crust toward the west, due to the resistance offered the advancing continental edges. This gave rise to the great mountain chains of the Rocky Mountains and the Andes. Along the western, or advancing, edge of the continents there would be a tendency to many minor adjustments of the floating crust. To the east of the Old Continent, however, there would be no pushing, but the eastern edge would drop off abruptly where it juts onto the heavier basaltic ocean floor. And in reality Japan and the Philippine Islands rest on the brink of a precipice, the deepest portions of the ocean adjoining them. On the edge of such a precipice we might expect sections to slide off, or other adjustments to take place frequently. Such disturbances undoubtedly give rise to the numerous earthquakes in the region mentioned.

Professor John F. Hayford of Northwestern University and Dr. Wm. Bowie, Chief of the Division of Geodesy in the U. S. Coast and Geodic Survey, have worked out certain facts that help us understand the cause of the rise and fall and buckling of the earth's crust. They have shown that were the earth's crust cut into blocks 100 miles square and 60 miles below sea level, the various blocks would weigh the same, irrespective of the fact that some containing mountains would have larger volume. They have proved this both by astronomical and by geodic calculations, and explain that unless such an equilibrium exists the Rocky Mountains would doubtless break down the terrestrial shell. They also point out that the lightening of a block as much as 3% would be sufficient to elevate the mass 9,000 feet. And as above explained, the lighter materials are close to the surface; for a weight of a cubic foot of earth at the surface is but 2½ times that of water, while the weight of the entire globe is 5½ times that of water.

With a few exceptions, such as the Alabama Hills in Inyo county, California, practically every portion of the globe at one time or another and usually numerous times, has been at the bottom of the sea. The silt and sand then deposited and later compressed into shales and sandstones, or other sedimentary rock may, after the region has become dry and lifted into a mountain chain, have been removed by erosion, leaving bare the granite mountain core. But land areas in general periodically rise as mountains, to be worn down by erosion and again form sea bottoms. This is due, not only to the possible westward drift of continents shoving up regions toward the front of their movements, and to the shrinkage of the earth causing the crust to become too large and thus wrinkle like the skin of a drying apple, but also to the constant shifting of weight of the land areas.

According to the ideas of Hayford and Bowie, above mentioned, as the mountains are worn down by erosion and carried into the sea, there is an increase of weight in the region where the material of erosion is deposited and a decrease of weight in the region from which it was removed. Now if the load on a raft is moved to one side, that side of the raft sinks and elevates the other side of the raft. Land areas are rafts of rock floating on a plastic ocean of basalt. But there is this difference between them and ordinary rafts, in that the region pushed down below some 60 miles depth becomes part of the flowage zone, melts off the bottom of the raft and moves to some region of lesser weight, there to push up some other section of the earth. According to the U. S. Geological Survey there is delivered into the seas and oceans from the United States alone 783 million tons of rock materials every year.

It will be seen that as the present mountains are due mostly to a shrinkage of some 200 miles in the diameter of the earth, that before such shrinkage there were probably no mountains. This is borne out by much evidence, and there is no doubt as time passes and new mountain ranges are formed that the new ones are larger than those of an earlier date. Throughout geologic time lands have gone down as well as up, but the sum of their movements have been upward, and the sea areas have gone up as well as down, but the sum of their movement has been downward. The land is gradually getting higher and the sea is gradually getting deeper.

At the commencement of geologic time it is thought our earth had an atmosphere similar to the present one except that there was very little oxygen in it. Although the oxidation of the rocks has consumed some oxygen, the influence of plant life has steadily been to increase the oxygen content of the atmosphere and make it more suitable for animal life. It has done this by utilizing the carbon and freeing the oxygen of the carbon dioxide gas in the earth's gaseous envelope.

This carbon dioxide gas is constantly replenished by volcanic activity. Volcanoes which are now thought to be due to local regions beneath the earth's crust becoming overheated through the activity of radioactive minerals, are not unmitigated evils as they are generally regarded. Instead of being vents through which the molten interior of the earth flows, they are vents for molten pockets of rock that have become intensely heated by radioactive minerals in particular regions. And they contribute carbon to the atmosphere. Carbon is one of the three fundamental materials at the basis of life, and were there double the life on earth that there is at the present time, all life would cease; for all the carbon in the atmosphere would be in the bodies of plants and animals, and death would overtake all. Furthermore, should volcanic activity cease it would not be long before the existence of life would be impossible because of lack of carbon.

The water also, so the newer geology teaches, came out of the earth as the earth cooled, through volcanic activity and warm springs. Much of it was added, there is good reason to believe, in later geologic time. Thus the waters of the ocean tend to encroach upon the land, even while the water falling from the sky wears down the mountains and carries them out to sea.

Two mountain ranges in North America, bigger than the Rockies, were lifted up and then worn down by such agencies before the present mountain ranges were lifted up. Sand and clay and mud are all the products of rock worn down by frost and wind and rain and glaciers. From the depth and size of such sedimentation can be calculated with much accuracy the size of the mountains required in their formation. The water constantly tends to wear down and deposit land areas in the sea, and Sir Archibald Geikie calculates that if the continents were thus deposited in the sea the sea level would be raised 650 feet, and if North America remained stationary half of it would be covered by the sea to a depth of several hundred feet.

Small warpings of the earth's crust are going on all the time, due to the shifting of the weight of areas through sedimentation. Such warpings usually elevate local regions only a few hundred feet. Erosion continues on a continent until the land area is but a little above sea level—a condition which has prevailed during most of geologic time—and then, largely due to astrological tensions, the crust yields to the strain of shifted weight and slowly, near the margins of the continents, folds and breaks in the formation of ranges of mountains from 1,000 to 1,500 miles long. These are called Minor Crustal Adjustments, and at least eight are known to have occurred in North America.

At still greater intervals—also largely determined by astrological conditions—there is a more complete adjustment of the land areas the world over. As the result huge mountain ranges are formed and the continents are elevated to a much greater height above ocean level. These are called Major Readjustments, and at least six are known to have occurred during geologic time.

The elevation of such masses of land has a decided effect upon the climate. New land areas change ocean currents, new mountain ranges change air currents, and even as now it is cold on a mountain top, so excessive elevation of land areas causes the climate to become so cold that the snow does not melt as fast as it falls. The mountains first become covered with glaciers, and these lowering the temperature of surrounding territory tend to spread the glaciers until a continent may be covered with an ice sheet from the north down to a latitude where melting takes place faster than the snow falls. Geologists know of several such periods in the past—each following a very long time of warm climatic conditions—when there were decided coolings of the climate, at least four of these periods being glacial.

A moment's reflection will reveal what a terrific effect such a change has upon life that has been living in a warm climate. A glacial winter lasting thousands of years causes the seas to deepen yet decreases their area, causing swift running torrents to flow where there was only sluggish water before, shutting off moist winds from the interior and turning that interior into desert, and in a dozen other ways upsets the conditions to which life has long become accustomed.

Origin of Physical Life

So far as known, at the present time all living things come from previously existing living things through some method of reproduction. In obedience to the second law of thermodynamics (the Carnot-Clausius law) inorganic evolution moves from the more complex to the simple. But life in its evolution follows the opposite course, and moves from the simple toward the more complex. Whether it had its origin here, or was carried to the earth from some remote sphere as a spore or seed embedded in a meteoritic fragment, material science has been quite unable to account for its beginning. Lecomte du Nouy in 1947 showed by probability calculations the inconsistency of believing the appearance of the first living cell to be due to a chance combination of inorganic molecules.

Psychical researchers have found, however, that whenever physical conditions were present that would permit the manifestation of intelligence, intelligence was always present there to manifest itself. Thus has it now been demonstrated that there is an inner plane, nonphysical in nature, and not subject to physical laws, where intelligences of incalculable grades at all times persist. It is on this plane that the unconscious mind or soul of man exists while it functions through his physical body, and it is on this plane that it will continue to function and develop after the dissolution of the physical vehicle.

As psychokinesis, the power of the mind to move and manipulate physical objects, has now been amply demonstrated in many university laboratories, and psychical researchers have observed the production of ectoplasm and materializations under the influence of the mind of a medium or some other entity, there is no valid reason to believe that under proper physical conditions influenced by suitable inner-plane weather, an inner-plane intelligence could not combine the necessary molecules to form a single cell organism, and continue to manifest through this primitive cell. Such "demonstrating" a physical vehicle by an inner-plane soul longing for physical experience would be no more remarkable than the occasional amazing "demonstration" by the psychokinetic power of the mind of health or some other physical condition that most of us have had opportunity to observe. The soul, launched on the Cycle of Necessity, as explained in Chapter 4 (Serial Lesson 4), Course 2, *Astrological Signatures*, has the power, now called psychokinesis, of attracting, molding and repelling the various forms that it needs for experience.

Whether it is the soul of a bacterium or the soul of man, it at all times resides on the inner plane. And so long as it manifests through a physical body it maintains its union with it through psychokinetic power. When, due to the stress of unfavorable inner-plane weather, which is mapped by astrology, external conditions offer sufficient resistance to the psychokinetic control of the body that it can no longer manipulate the physical functions and handle the electromagnetism—which is the boundary line energy that links its high velocity to the low velocity of the physical—we say the body is dead.

That, however, merely signifies that it has lost this particular physical vehicle. It still persists on the inner plane, and if it is a form of life lower than man, it strives to make contact with the germ of another and somewhat more complex physical life form through the physical existence of which, as it grows to maturity and perhaps to old age, it can gain still other physical experiences.

The soul has two faculties that have now been amply demonstrated in university laboratories. It has the faculty of acquiring information through what is now called extrasensory perception. It can, without the aid of physical senses or reason acquire information about the distant present, the past and the future. Clairvoyance, telepathy, post cognition and precognition have now been thoroughly demonstrated. Extrasensory perception, including telepathy, is the normal manner in which information is acquired by inner-plane entities. But when an inner-plane entity forms a union with a physical body, it largely focuses its extrasensory power on the responsiveness of the physical organism and acquires most of its information from what happens to the physical body.

By the time it has acquired experience enough to be able to become united to a human being it depends very largely for its information on the sense organs and the brain. It retains its ability to get information other than through the nervous system; for it has been shown that while they may be objectively unaware of the information thus acquired, both man and animals often act successfully to adapt themselves to approaching situations of which they could have no knowledge through physical channels.

While its physical body lives, however, the soul of any life form, including man, uses its extrasensory faculty chiefly to keep aware of what happens to that physical body. It is united with that body to get physical experience, and its extrasensory faculty is chiefly concentrated on becoming aware of physical experiences, which in man, of course, include the electromagnetic processes within the brain that give rise to objective consciousness.

As university experiments also demonstrate, when a soul unites with a physical organism for the purpose of material experience, it does not lose its power of psychokinesis. It can still, on occasions, move and manipulate physical objects without physical contact. But while thus united to a physical body, its psychokinetic power is chiefly exercised in keeping in contact with that body, and controlling its movements. Every voluntary movement made by man is due to the soul on the inner plane exercising psychokinetic power over his motor nervous system.

What has here been stated also explains why when people exert themselves to get information through extrasensory perception they usually fail. The willing effort has been conditioned throughout a long past to concentrate the extrasensory faculties on the reports of the physical senses. What is needed, instead, is a strong desire on the part of the soul, or unconscious mind, rather than of objective consciousness, to get the information more directly without the intermediation of the physical senses.

And it also explains why intense willing usually thwarts the effort of those who try to demonstrate something through mental power. The willing process has been conditioned throughout a long past to concentrate the psychokinetic power on moving the muscles to accomplish what was desired. But what is needed is a strong desire on the part of the soul, rather than of objective consciousness, to accomplish the demonstration more directly without the intermediation of physical movement.

Other factors, which will be considered in subsequent chapters, play a part in the evolution of life forms on the earth; but certainly the power of the soul to move and manipulate physical substance is a highly important factor in this progress from the simple to the more complex.

The First Life on Earth

Apparently as soon as the earth had sufficiently cooled and other conditions developed that made it possible for life to function here, about 1750 million years ago, inner-plane life succeeded in using its psychokinetic ability to get a foothold on this sphere. At that time the temperature probably was considerably higher than now, there was little free oxygen in the atmosphere, and sunlight was shut off by dense clouds. It is estimated that there is in the sedimentary rocks and in the fuel deposits of the earth, 30,000 times as much carbon as there is at present in the atmosphere. Higher forms of life could not live under such conditions as doubtless existed when all this carbon as carbon dioxide was in the atmosphere. Together with water vapor it must have formed an atmospheric blanket that absorbed the rays of the sun and kept the heat of the earth from radiating. Under these conditions bacteria, the lowest form of life of which we have any knowledge—although Coenocytes, Mycetocytes, certain molds and certain algae are amorphous living matter not divided into cells, while bacteria are unicellular but lacking in definite nucleus—would thrive and prosper.

Plants are dependent upon light for the assimilation of the carbon dioxide of the air, which is their chief and most essential food supply. The nitrogen bacteria have the power of assimilating free nitrogen from the air and at the same time and without the aid of sunlight can decompose carbon dioxide. They thus can live on inorganic products without the aid of sunlight, which plants are incapable of doing.

All life on earth—bacteria, plants, animals, and man—is associated with protoplasm. The four most important elements in protoplasm are nitrogen, carbon, oxygen and hydrogen, which the primitive bacteria obtained from the free nitrogen and carbon dioxide, and the water, of the air.

Among the oldest rocks of the earth formed after the process of erosion set in and conditions developed that made it possible for life to gain a foothold—estimated by the most competent authorities as 1750 million years ago—at the commencement of the Archeozoic era, are to be found immense deposits of mineral that have been formed by bacteria which have developed from the simpler form. Iron bacteria, such as Lepothrix, obtain their energy from the oxidation of iron compounds. The iron oxide so obtained being insoluble, stays in the bacteria, and when the bacteria die this iron oxide remains as a mineral deposit. Vast beds of iron ore formed in this manner are known. Sulphur bacteria in a similar manner oxidize hydrogen sulphide, and the remains of their dead bodies form huge ancient mineral deposits.

Plant Evolution

From its first appearance on earth life possessed and expressed the three hereditary drives (Chapter 5, Serial Lesson 60, Course 5, *Esoteric Psychology*)—the drive for significance, the drive for reproduction, and the drive for nutrition—which are the most powerful motives in human life. From the very first there is exhibited in the effort to develop new and more complex forms not required for survival—for iron bacteria and innumerable other forms of life still exist abundantly in practically the same condition that their ancestors existed when their earliest remains were deposited in ancient rocks—the drive for significance, the drive to ascend to something better. And in so doing not only did some individuals develop more complex structures, but the direction of movement of all such forms as are the ancestors of existing life forms on earth, was toward the fulfillment of God's Great Plan (Chapter 1, Serial Lesson 56, Course 5, *Esoteric Psychology*). It is true that inner-plane and outer-plane environmental changes at times forced certain life forms to alter structure and habits or perish. But there is evidence also of the drive for significance satisfied by more complex experience.

Animals are entirely dependent upon organic food for their existence, for they are not provided with chlorophyll. This organic food is supplied by plants. There are some plants—mushrooms, molds, mildews and rusts, as well as certain flowering plants—that have no chlorophyll. They must depend upon the organic food which has been gathered—that in some cases has decayed through the action of bacteria and in others yet exists in the living plant—for their food supply.

While some of the bacteria in the world today are injurious to mankind, through leaving their by products where they poison him, yet organic life is dependent upon bacteria for continued existence. Bacteria not only assimilate free nitrogen, and change certain nitrogen containing substances in the soil into forms that can be used by higher plants, but they bring about the decomposition of dead organic material, which is essential if it is to be used by plants. All organisms give off waste products, but with the exception of carbon dioxide little of this waste matter can be used by plants until it has been decomposed, or rotted, through the action of bacteria.

The nutritive liquid of animals is blood, and this in the higher animals contains the red pigment hemoglobin. It is chemically quite similar to the nutritive liquid of plants, but differs from the latter in that a molecule of hemoglobin contains one atom of iron, whereas the simpler molecule of chlorophyll contains one atom of magnesium. And there are certain lower type animals, including some snails, in which the blood molecule, instead of either iron or magnesium, contains an atom of copper.

There are also elementary forms of life that seem to have the outstanding characteristics of both animals and plants. The dinoflagellates are algae so small as to be visible only under the microscope. They are abundant in stagnant water. They move quickly through the water by the aid of long flexible tails, and as they breathe their cellular body inflates and deflates. In these motions they resemble animals. But they resemble plants in that they are single celled organisms containing chlorophyll and surrounded, as are many plants, by a cellulose membrane.

Life was not content with such simple existence as the bacteria. There was the urge for more complex expression. Under the stimulus of inner-plane and outer-plane environment its psychokinetic power produced alterations in some individuals which were transmitted through cell division. A very elementary plant appeared, the blue green alga, which still exists today. In some of these, which in form and reproduction resemble bacteria, there is no nucleus and no chlorophyll. Their pigment is phycocyanin.

Certain of the blue green algae occur as slimy blackish green films. They, like bacteria, reproduce by simple cell division. Some of them, similar to bacteria, are able to endure heat that would be fatal to ordinary plants. The sinter deposits, or formation, of the hot springs and geysers in Yellowstone Park are due to such algae. So also in cooler water the presence of a free floating form of blue green alga, so called but in this case red, gives the Red Sea its characteristic color.

The next step was the development of a nucleus in the cell and the ability to manufacture chlorophyll. Green plants are able to use chemical elements in such proportions as to manufacture the substance which gives to leaves their green color. This chlorophyll, in the presence of sunlight has the property of capturing carbon from the carbon dioxide in the atmosphere and releasing free oxygen. The process by which it does this is similar to that by which sunlight causes chemical changes to take place on a photographic negative, and is called photosynthesis.

The early plant, such as the green algae, consisting of but a single cell, needed a certain amount of protection, and this desire directed psychokinesis to the formation of a cellulose wall about the protoplasm within. The protoplasm in all but the very lowest plants, even as is true in all animal cells, contains a well-organized nucleus. Growth in plants and animals alike takes place through cell division in which both nucleus and the cytoplasm—the protoplasm of the cell exclusive of the nucleus—split, a portion of each going into the production of new cells. The protoplasm in the body of man today may contain an infinitely small amount of the protoplasm of a primitive one celled form of life that existed more than a billion years ago; for—even though psychokinesis was necessary to form the first primitive cell—so far as has been observed new cells are formed only by the division of cells previously existing.

Even as today it is necessary that nations cooperate, so it is evident that in many cases groups of cells could gain an advantage by cooperating. This need, recognized by the soul's extrasensory perception, was met through the psychokinetic power of its desire. As a result we find the next step in progress to be, instead of single celled plants, plants composed of a number of cells. The simplest of these are the filamentous algae, consisting of rows of cells somewhat like a chain, barely attached to one another. When such a colony of cells finally became established, the next step would be toward a division of labor, and we find a tendency in somewhat higher forms of algae for certain cells to specialize in gathering carbon from the air, and others to specialize in the storage of the food so gathered, and still others in protecting it from the evaporation of its water and the inclemencies of its environment.

Before we pass to the next step in the development of plants of more than one cell, let us pause a moment in awe before the vast work of the primitive algae that early in the geologic history of the world must literally have swarmed the seas. We are somewhat familiar with coral polyps, minute colonial animals which build islands and shore lines with their dead bodies; but such land building does not compare in its extent and importance with that of certain lime secreting algae. These calcarious algae, as they are called, are held to be responsible for the formation of the very ancient limestones. The rocks of the Grenville Series alone, a very ancient series of rocks, are nearly 18 miles thick, and half of this is limestone undoubtedly deposited by such algae. In other cases the algae and a lime secreting bacterium are jointly responsible, as in the case of the massive limestones of the Teton region.

This habit of secreting lime, which was later adopted by animal life, has a most important bearing upon any study of the past; for before this neither animals nor plants had hard parts that could be preserved as fossils in the rocks, and their presence can only be known from inference. Such an inference as to the extent of ancient life on the earth may be found in beds of iron ore and sulphur as previously mentioned, and in the existence of masses of graphyte in exceedingly ancient formation. Graphyte is never produced in nature other than through organic activity.

Most of us are familiar with green "pond scums," which are chains of algae cells, all quite alike, floating on the water. These are fresh water algae, but certain kinds that have developed from them, and become more elaborate in structure have found their way to the sea and form the green sea weeds; and others, because they secrete lime, look very much like plant corals. The bulk of marine vegetation or seaweeds, however, have developed other traits to suit their salt water environment and belong either to the Brown Algae or the Red Algae. The red algae, which constitute the greater bulk of seaweed, is thought to be but a more complex development from green algae. The red pigment and the brown pigment, by which the red algae and the brown algae are colored, is supposed to supplement the action of the chlorophyll in utilizing the light which filters to it through the water.

The brown algae, including the giant kelps which are so common to the Pacific Coast of America and so familiar to those who visit the ocean beaches near Los Angeles, sometimes reach a length of one hundred yards or more. They are probably not direct descendants from green algae, but from the animal-like Flagellates, to which group the previously mentioned dinoflagellates belong.

The ancient seas were fresh, for the salt now in the sea was gradually leached out of the land. The adaptation of life to salt water, then, is of a later date than the more ancient rocks that have formed by sedimentation. The giant kelps and the red algae have solved the problem of living in salt water better than any other plants, and seem to have reached a point, due to the restrictions of their environment, beyond which further progress is impossible. Their texture is such, due to the manner in which the cells join, that while immersed in water it freely circulates through them, yet the outer cells have been thickened and toughened to form a leathery skin which, when exposed to sun and air, due to low tides, protects it from evaporation. Some of them, like the giant kelps, have developed not only an anchoring device, called a holdfast, by which one end is attached to a pebble at the sea bottom, but also hollow bladder-like buoys that may be as large as a child's head, by which their long stems, bearing floating leaves, may be made to reach the surface.

It is supposed by some naturalists that the fungi are descended from certain species of red algae. The fungi do not possess chlorophyll and depend upon other plants and animals to furnish their carbon food supply. They probably have degenerated from higher plant forms, finding an easy living at the expense of others. Parasitism, whether in plants, in animals, or if we may use the term thus, in man, is always followed by deterioration. These fungi—the molds, mildews, rusts, mushrooms, etc.—some 40,000 species of which are known—have degenerated to a very low level in plant life. They do not possess seeds, but propagate by means of spores. The smoke that issues from a puff ball when pressed consists of millions of such spores. In the case of the familiar mushrooms and toadstools the spores are developed in the gills on the under side. In fact these gills, or flutings, open for the express purpose of dropping the myriad minute spores by which they reproduce.

Fresh water ponds are in the habit of drying up. In such instances the green algae living upon their surfaces, unless possessed of some method of tiding over the dry spell, all die. The mud at the bottom of such a pond, when the pond first dries, is moist, and the algae would cling to it for moisture, for active life either in plants or animals depends upon the protoplasm being supplied with moisture. The water gives to protoplasm a semi fluid consistency which is absolutely essential to its movement.

Green algae, resting upon the drying bottom of a pond, would be hard pressed to prevent all its moisture being dried out by the sun, and to get an additional supply from the drying mud. The desire for life in some of these brought psychokinesis to bear to make necessary structural changes. The algae, by a thickening of the cell walls, escaped being completely dried up, and thus when the dry spell was over was able to resume normal life. These special thick walled cells, which foreshadowed the development of seeds, are called resting spores. Some of the algae also, in their desire to follow the water as it receded into the mud and thus provide themselves with moisture, developed cells in the direction of the moisture, and these cells becoming specialized were the first roots. This was one of the greatest and most important steps taken by life since it started on our globe, for it gave rise to the ability of life to live upon the land.

Vegetable life in the water depends upon the water for support, but as life crept from the warm and shallow ponds and fresh water seas out upon the land it found it to be a great advantage to be able to lift its chlorophyll bearing surface to the sun, that it might draw a greater food supply from the air. Some of the liverworts, which lie prostrate upon the ground, have delicate hair-like roots, and a structure not as complex as the algae, being composed of almost uniform cells. They live today as examples of what the first land plants must have been like. But with the desire strong upon them to reach the light, certain of the cells developed a harder, more compact structure, and gradually a supporting stem came into being.

With the development of firm supporting tissues the need arose also for special tissues for the rapid transportation of water, and a softer conducting tissue was developed. Not only are the liverworts prostrate, but so are some of the other low land plants such as the mosses. And to indicate that their ancestors came from the water we find that mosses and ferns are dependent upon the presence of free water for the development of certain phases of their life histories. Even as amphibious animals must return to the water to lay their eggs, and pass through the early stages of life in water, so familiar to us in the lives of frogs and toads as the tadpole stage, so these plants also may be considered amphibious.

The ferns, although reproducing by means of spores instead of seeds, are more complex in structure and in their life histories than the liverworts and mosses. A spore is a single cell, minute in size and without sex, and in the case of the fern a number are born in each little capsule on the under side of the frond (leaf). When this spore is released and germinates it does not grow into a fern, but into a very different plant, or prothallus, a green blade about a quarter of an inch long. On the lower side of this new plant grow the sexual reproductive organs which produce the egg cells and sperms. The sperm has a tail of minute hair-like cells by which it swims through the film of water that must be present on the blade of the plant, to the egg, which it enters and impregnates. And from this the new fern grows. In the case of the mosses the generation that produces the sexual parts is the moss plant, the other plant essential in the life cycle being the capsule which bears the spores. This lives as a parasite on its parent. Spores are not seeds, but they serve as resting bodies through which later a generation may be perpetuated, and they serve as a convenient means for distributing the species.

Somewhat more complex than, yet evidently related to, the ferns are the curious horsetails that grow in low moist ground. Some twenty-five species are known to exist at present, representing, in a meager way, the gigantic species that once existed upon the earth before the advent of flowering plants. The club mosses, also spore bearing plants, are supposed to be related remotely to the ferns, and once provided an important part of the land vegetation.

The dependence upon water for the propagation of the species became a serious handicap to land plants, just as it did to land animals, and the problem was solved much in the same way by both. In the case of a seed plant the pollen falling on the ovule develops a little tube that penetrates the egg and brings the fusion of the male and female elements that are necessary for the beginning of a new plant. This does not require the presence of water through which the sperm must swim, and has an additional advantage in that the young plant resulting from the fusion of male and female elements remains associated with the parent plant, drawing nourishment from it, and protected from inclemencies by being enclosed in a sheath and surrounded by nourishing food. When the little plant in the seed reaches a certain stage of growth its development is stopped for the time being, to begin again when the plant has left its parent and found its way into moist soil.

Seed bearing plants release their young alive, quite as effectively as do the higher animals. The young plant, or embryo, which can clearly be seen by opening a soaked pea or bean, has another great advantage over the plants growing from the sexual union of the parts that grow from a spore. The latter must procure all their own nourishment from the start. But seed plants have an abundance of food stored up in the seed to give them a good start on life's journey. They are as well provided for as the calf which grows inside its mother from an egg to considerable size before being born, and then after birth is provided with rich warm milk for six months or more. The seed plants take excellent care of their young.

The first seed plants were ferns, now extinct, but existing in great numbers during the Paleozoic era which began about 350 million years ago. These seeds were less perfect than those of today, and no fern now exists that bears seeds. The cycads and ginkgoes, once very numerous upon the earth, are clearly descended from ferns, and represent no great modification in structure. The "sago palm" of our greenhouses is one of the cycads, and the Ginkgo, or maiden hair tree, is quite common as an ornamental tree here in California.

It is thought that the conifer, or cone-bearing trees, are modifications of certain club mosses whose fossil remains have been discovered. A small species of club moss is common on the hills of Los Angeles. The cones of conifers seem to be mere modifications of structure common to certain extinct club mosses which are known to have borne seeds. These cone-bearing plants, represented most extensively by our pines and firs, are of a lower order of existence than most of our flowering trees and plants.

The thin, long, resinous foliage of our conifers is an adaptation to prevent the excessive evaporation of moisture from the plant in dry regions. Other plants of the same group, such as the Araucaria which is common in California parks, have broader leaves. These trees came into existence upon the face of the earth at an earlier date than the common flowering plants. The seed, instead of being enclosed in an ovary, is naked like those of the cycads, and is borne on the surface of a scale. These scales, bearing naked seeds on their surfaces, form the cones of familiar trees.

Even though the conifers came into existence so long ago, they have proved exceedingly successful, as our vast northern forests prove. Some of them, too, have developed an uncanny way of anticipating the future, as in the case of the fire type pines, which hold their seeds for a dozen years until a fire destroys all other vegetation, and then, due to the heat that has passed, the cones gradually open and the seeds are deposited in the ground that has been well prepared for their being covered, and from which the competition of other growth has for the time being been eliminated.

Chapter 3

Progress of Invertebrate Life

Serial Lesson Number 127
Original Copyright, 1949
Elbert Benjamine
a.k.a. C. C. Zain

DUKE, the larger dog of Mr. and Mrs. Elbert Benjamine, was born Sept. 29, 1938, 11:00 a.m., P.S.T. 118:15 W. 34:03N. Birth precisely timed. He has had only two dog friends, Pomeranians, both now dead. The smaller dog in the picture is the first of these friends for whom he showed great affection. On the dog level Duke has responded to both major- and minor-progressed aspects as do men. It was easy to anticipate the time and kind of ill health, good times and bad times, and the events that would come into his life.

1939, April 6, eye trouble: Mars inconjunct Saturn p.

1939, April 22, new home: Venus sextile Neptune r.

1939, May 22, injured by auto: Mars inconjunct Saturn p.

1939, July 14, long journey: Venus sextile Neptune r.

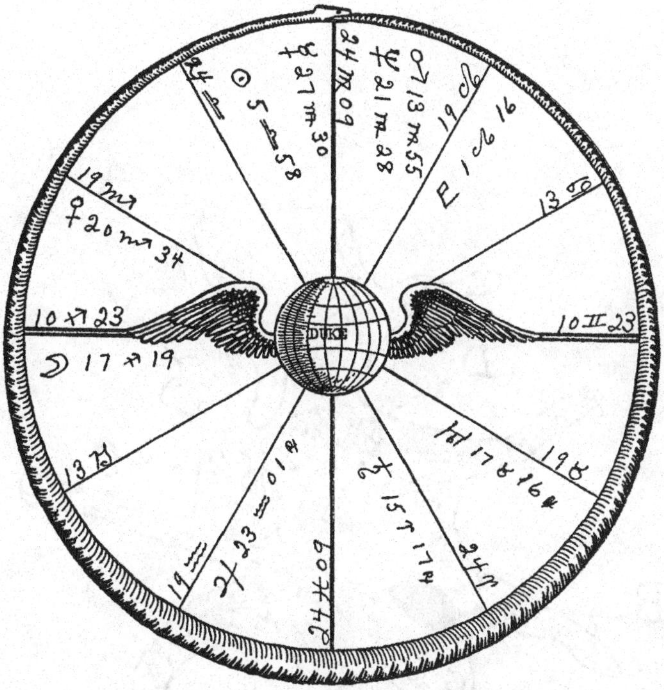

DUKE

1942, Nov. 2, new home: Venus square Jupiter *r.*

1943, Aug. 2, vacation in mountains: Mars trine Uranus *r.*

1945, Aug. 22, master taken to the hospital; refused to eat until master started to improve: Mercury semisquare Venus *r.*

1946, Nov. 28, dog friend died: Mercury semisquare Venus *r.*

1947, Aug. 22, long auto trip: Mercury semisextile Mars *p.*

1948, March 6, operation for fibroid tumor on right elbow: Sun opposition Saturn *r.*

1948, April 27, operation for fibroid tumor on left elbow: Sun opposition Saturn *r.*

1948, May 4, auto trip to Key West, Florida: Venus sextile Mercury *r.*

1948, Aug 23, lost molar tooth: Sun opposition Saturn *r.*

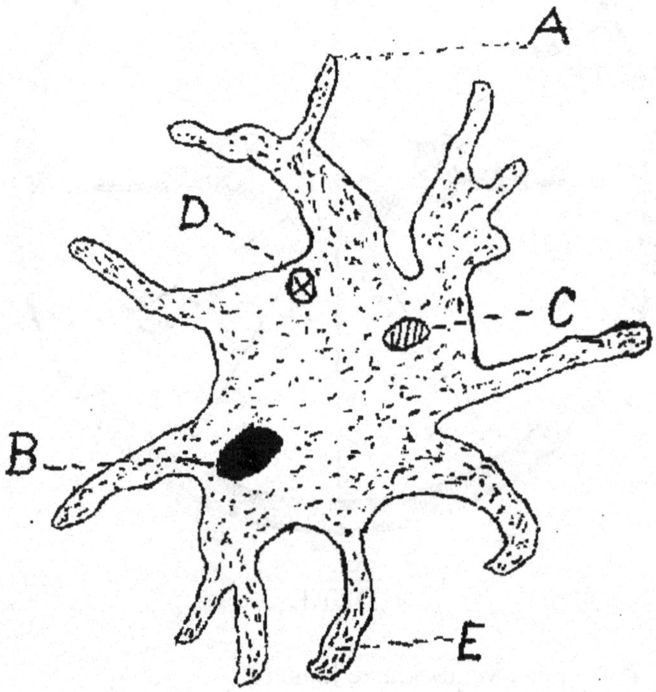

Diagram of an Amoeba

The amoeba is a typical Protozoon. It appears as an irregular speck of greyish jelly about 1/100 of an inch in diameter. It is common in fresh water ponds, where it oozes along engulfing other tiny specks of organic matter by flowing over them. "A" represents one of the outflowing lobes that surround the prey. "B" is the nucleus. "C" represents food that has been ingested. "D" represents undigestible pieces of food that are about to be expelled. "E" points to the granular struture of the protoplasm.

Chapter 3

Progress of Invertebrate Life

WHEN the environment radically changes, most life forms must also radically alter either their structure or their habits or perish. It will be indicated in subsequent chapters how marked weather changes brought the extinction of numberless forms, and the kind of radical changes some made in order to adapt themselves sufficiently to the new condition to be able to survive. When previously arid regions became deluged with rain, their vegetation had to acquire new characteristics to prosper, and when previously well watered regions became arid, not only did plants have to change their structure to gain and preserve the scant moisture, but creatures whose young were hatched in the water and developed through the tadpole period had to so change that their young reached a land inhabiting stage without living their youth in the water.

In a previously calm region, when climatic changes developed powerful winds, both plants and animals, to survive, had to develop protection against such gales. When glaciers came down from the north and the previously warm weather became intensely cold, some animal forms developed feathers from scales, and others developed fur; and the young, to protect them from such inclemencies were either not born in the previously immature condition, or the eggs and young were protected by nests until the offspring had matured enough to be able to adapt themselves to the food and temperature conditions brought about by the severe weather.

Wind and falling water and the cold which causes the moisture in its capillaries and crevices to freeze and crack open or flake off pieces of rock, have cut canyons thousands of feet deep in the earth's crust and have leveled giant mountain chains.

These facts are familiar to all. But there is now good evidence that the inner-plane weather, consisting of astrological energies, is fully as powerful to inaugurate changes in the rocks, and in life forms, as is the outer-plane weather. Even though they are not aware it is influencing them, Church of Light research has proved that inanimate objects and other life forms as well as man are powerfully thus influenced. A machine built at one time will not last as long as a machine built at a more favorable astrological time. Crops, as many farmers have learned through experience, even though planted when the outer-plane weather is favorable, will not thrive and produce so well if planted when the inner-plane weather is adverse as when planted while inner-plane weather is more suitable.

So far as our research has gone we have found that progressed aspects affect animals in the same manner they affect men, due allowance being made for the normal level of the animal. It is not to be expected that an insect whose normal life span is only a few days will respond other than to progressed positions within those few days. And just how they so respond is yet to be ascertained by timing their births and calculating the aspects which subsequently form. But we have had ample opportunity to observe in cats, horses and dogs that on their level progressed aspects influence their lives as they do human beings.

A lady in San Francisco who raised show dogs kept a record of each puppy born, erected its birth chart and observed its progressed aspects. These charts were sent to us for our opinion. If one of her dogs had a good tenth house it had a good chance of winning a prize at the dog shows, and she found that this was almost certain to occur if it had a harmonious progressed aspect to the ruler of the tenth at the time of the show. The same kind of chart and progressions that belonging to a politician would insure he would win an election, if it belonged to one of her dogs would equally insure he would get a prize at the dog show.

Another gentleman in Montana used astrology in raising and training race horses. He learned to tell when the colt was first born what its prospects were, whether it would be subject to accident, and if it were worth training as a racer. And he could tell by its progressed aspects at the time of a given race whether the horse would get a good break, or whether he would have to overcome fortuitous obstacles in order to win.

The writer has watched the response of both cats and dogs to progressed aspects. The chart of his present dog with the dates of all-important events that have entered his life during more than ten years is given on page 42, together with the significant major progressed aspects coincident with them. Due to lack of space the minor progressed aspects and transit progressed aspects are not there given; but anyone who wishes to calculate them will find that exactly as in a human horoscope, each of these major progressed aspects is reinforced by a minor progressed aspect, and released by a transit progressed aspect, within one degree of perfect at the time the event occurred.

Life must adapt itself to both inner-plane and outer-plane weather or perish. And the survival of the fittest is a factor in organic evolution. But as subsequently will be indicated, such progress is not in some haphazard direction. Every outdoor naturalist I have ever met has been convinced there is a Super Intelligence permeating and broadly directing all the processes of nature. Such direction is not that of whim or prejudice, but always according to well-defined laws. And one of the outstanding laws is that the pressure of the inner-plane weather and the outer-plane weather is such that the overall progress made by life—even though innumerable forms do move up blind alleys and become extinct, and others finding satisfactory adaptation stagnate—is toward filling in the universal plan formulated in the mind of Deity.

Plants as well as animals and men have, in some degree extrasensory perception, and psychokinetic power (Chapter 2, Serial Lesson 40, Course 1, *Laws of Occultism*). And in response to their desire to live and find an adequate food supply they have made many remarkable and intelligent variations. Our stone crops, for instance, finding competition unusually strenuous on fertile ground, gradually moved into rocky regions where other plants did not grow. In such ground moisture is retained but a short time. Therefore, to meet this condition the stone crop greatly thickened its leaves, so as to make a reservoir for holding the moisture and thus tiding it over dry weather. The various species of cacti, finding a desert environment developing around them, likewise thickened their leaves as water reservoirs. In addition they had to combat a scorching sun and numerous herbivorous animals, made voracious because other vegetation, always scanty, failed to grow during the long dry seasons. To meet the scorching rays of the sun they caused the outer cells of their leaves to harden, thus coating each leafy reservoir with horn-like insulation against the evaporation of its water. To protect themselves from greedy animals many of their leaf parts were made slender and hard, so that their leaves were covered with thorns.

Botanists recognize the leaf as the basic form of all the organs of higher plants. However diverse in form and function a plant organ may be—bud, thorn, flower part, bulb, or fruit—it is but a modification of leaves. In the calyx of the peony, for instance, the sepals, while largely green like any other leaf, have a fringe of color, indicating the process of transformation. This change of leaf into petal has not been completed in the snowflake; for here we find the petal of the flower white, except the very tip, which is yet green like the leaf. In the begonia, also certain of the stamens often revert to their original leaf form; and in the water lily the stamens and petals grade into each other with such slight variations that it is easy to trace all the steps of enlargement, broadening and coloring, by which the leaf-like stamen becomes the beautiful petal. Thorns and the stings of nettles are also mere modifications of leaf structure in answer to the intense desire of the soul of the plant to be protected from its numerous enemies. And even as the most delicate rose, or the most gorgeous orchid, results from modifications of leaves, so every animal on the face of the earth is but the result of modification of simple single celled protozoa.

Plants growing like the water lily, where there was little competition for sunlight, developed broad leaves. Those growing where there was much competition for sunlight, like our grasses, developed narrow leaves that were able to profit by whatever gleam of light filtered through the surrounding vegetation. We find, in fact, much the same tactics employed by plants that are employed by animals for the same purpose. Plants produce poisonous and evil smelling secretions to ward off enemies, much as do certain ants and beetles among insects, and as does the skunk among mammals. Some plants also are carnivorous. The sundews, the butterworts, the bladderworts, the Venus fly trap, and the pitcher plants—one of which grows in the mountains of California—all trap and assimilate insects.

In the Venus fly trap there is a rounded blade. On the upper surface of each half of this blade are three prominent bristles, and around the margin a row of stiff thorn-like teeth. When an insect touches one of the bristles there is an electrical charge in the plant similar to that taking place in an animal when it contracts a muscle, and the two halves of the blade clap together the marginal thorns interlocking like the teeth of a rat trap. Then a digestive fluid is secreted and the insect so caught is digested and assimilated, after which the blade opens for another capture. It may be cheated by using a little piece of moist paper to take the place of an insect, but after twice closing on worthless material in rapid succession it usually will refuse to be duped a third time. It modifies its actions because there is *memory of a previous experience*.

It would be interesting to write a large volume citing the marvelous methods plants use to overcome the difficulties that have confronted them. It must suffice here, however, to say that every plant form and method of life holds the story of its endeavor to overcome certain limitations placed upon it by environment. The deciphering of these plant romances and adventures, as well as those of insects and other animals, has been my chief and pleasantest avocation for more than half a century, but they cannot be related here.

Early Animals

Living matter is always associated with protoplasm. Protoplasm is an essential ingredient both of animals and of vegetables. Where, then, is the line of demarcation between them, and what were the incentives that produced the first animal?

As I have pointed out in Chapter 2 (Serial Lesson 126) there is no clear cut line between them; some animals, such as the protozoan Luglena, are provided with chlorophyll, and others, such as the ascidians, possess cellulose; both of which commonly are considered strictly plant features. Animals live upon organic matter, and in some stage of life possess the power of locomotion. Yet among plants the fungi live upon organic matter; and many algae, such as the diatoms, and the spores of the cryptograms, have the power of locomotion. To be sure, the male sexual element of most plants has the power of locomotion well developed.

In general, the source of food supply and the power of locomotion tend to distinguish animals from plants. Plants, with the exception of those that feed upon material already organized, possess the green coloring matter chlorophyll, by which, in the presence of sunlight, they are able to capture carbon, their chief food supply, from the atmosphere. Animals, on the other hand, are not capable of living upon inorganic matter. Their chief food supply is the organic matter stored up by plants. Animals also feed upon other animals. In fact, sea creatures form a chain from the smallest to the largest, the smaller in turn being devoured by the larger. But the original food supply sustaining the smallest, and hence the whole chain, is vegetable or bacterial in origin.

To obtain a vegetable food supply, either the plants must be brought to the animal or the animal must go to the plants. Water tends to bring the food supply of certain creatures, such as the sponges, yet even these usually have developed the power of producing the current of water which brings their food. But more often, to get an adequate food supply, the animal must go to its food. This necessitates locomotion.

We can hardly conceive of animals living before plants or bacteria, but so soon as these came into existence there was an available food supply, and it is probable that it was not long before there were animals developed to take advantage of it. In fact, it is even possible that animals developed before plants, as many of the protozoa feed freely upon bacteria, and today thus exert a limiting influence upon bacterial activity.

The urge to secure a food supply—the drive for nutrition—is a fundamental impulse common to all life. And this intense desire ever tends through psychokinesis to adapt the structure to the end of better securing its food. When a new condition arises the soul through extrasensory perception is dimly aware of its plight and feels the desire successfully to meet the new condition. Psychokinesis endeavors to provide the way, but subjective intelligence makes many mistakes. It is not reason, but the primitive working of psychokinesis based on imperfect perception and the memory of previous experiences, which may have been largely astral, stored in the astral form.

But desire changes the astral form and this in turn through psychokinesis changes the form and attributes of the physical structure. Thus we may conceive of a single primitive cell of living matter, stimulated by desire for food, departing from the custom of seeking nourishment from the inorganic matter and appropriating the food already secured by its neighbor. This then proved so successful an expedient that the cell adopted it, and when it divided to form two cells, each new cell continued the trait. But this method, to prove permanently successful, requires that the cell be able to move from place to place in search of other cells to devour. This desire actively to seek a food supply, through psychokinesis brought about a change in the physical structure that gave greater mobility and finally resulted in a cell having the power to move about ingesting less favored forms of life. Such was the primitive protozoan.

The protozoa not only were the first animals on earth, but persist today as the most abundant aquatic animals. Millions of them swarm in almost every drop of water. Not all of them are so small, however, for they range from those microscopic forms just mentioned up to a gigantic species two thirds of an inch in length found as a parasite in the intestines of lobsters. They are all single-celled creatures.

Animal life is divided by naturalists into twelve great groups, or phyla. Unfortunately, knowledge of animal life is so greatly confined to the very few that there are no vernacular names for most of the great groups of animals living today. This is true with even greater force of extinct animals, of which I shall speak in treating of mammals. Consequently, while I desire to avoid technical names, I must be pardoned for occasionally using them in these lessons, because there are no other terms by which a great number of interesting creatures may be designated.

PHYLUM I, the Protozoa: These are infinite in the variety of their forms. The typical protozoa is the amoebae (page 44), which is abundant at the bottom of fresh water ponds and among decaying water vegetation. It is a microscopic mass of jelly-like protoplasm containing a nucleus. It moves by changing the outline of its body, pushing out and withdrawing portions of the jelly-like mass to produce a flowing effect. Its food consists of minute animals or plants or other bits of organic matter. When it touches such a morsel it gradually flows around and over it until the latter is quite surrounded. The protoplasm surrounding the food particle then secretes an acid which kills the prey and forms the soluble peptones or digestive ferment necessary for digestion. When the digestible portion of the food has been assimilated, the undigested particles are left behind as the amoebae flows on.

Such a simple organism is removed from certain primitive single celled plants only by a slight modification, for we must remember that some of these plants have the power of locomotion. Certain plants also feed upon organic matter. The protozoa, therefore, but utilize in a somewhat greater degree of coordination, two principles that also are used by plant forms. We may assume that the frothy chemical compound called protoplasm found it more expedient in the case of the protozoa to flow slowly about feedings on particles of life that had been already organized than to remain in one place and endeavor to transform inorganic elements into food value.

Yet because of its minuteness and simplicity of structure we should not hastily scorn the simple cell. The single celled protozoa have an infinite variety of modifications, and the cells that make up the body of both plants and animals are not widely dissimilar to these. Were it not, for instance, for the amoebae-like cells in the human blood, man would soon succumb to infectious diseases. The white corpuscles of the human blood often are called amoeboid corpuscles, because to all intents and purposes they are amoebae cells belonging to the human organism that are fostered by it as soldiers to guard it against invading germs. The amoeboid corpuscles, when minute organisms of various kinds invade the human system, act toward them as the ordinary amoebae act toward their prey. They pursue them and flow over them, engulfing them in their protoplasm. They are then digested and portions not assimilated are carried by the blood stream to parts of the body where they may easily be expelled. It is only when microbes multiply to such an extent that they so outnumber the amoeboid blood cells that these cannot kill and devour them that such diseases prove fatal.

I have mentioned in Chapter 2 (Serial Lesson 126) that certain algae devised the expedient of secreting lime. Other early plants—such as the microscopic ones called diatoms, closely related to algae and supposed to be the source of the oil in Southern California oil fields—adopted the expedient of secreting a skeleton of silica. So we need not be surprised that early one celled animals also should secrete hard parts to protect themselves from other predatory one celled animals. Certain of the protozoa, called foraminifera, secrete a shell, or external skeleton, of lime. There are foraminifera also that secrete a covering of chitin. Chitin is the horny substance forming the outer coat of insects and the crayfish group. Others of the protozoa secrete an external skeleton of silica. We see, therefore, that among the very primitive single celled organisms of both plants and animals there existed not merely the power of nutrition and reproduction, but also the power to secrete substances that were not protoplasm.

This is very important to us; for man's body, like all organic forms, is built up by cells. The skin and viscera, in fact, consist of cells. But the bones and muscles are chiefly the secretory products of cell activity which continues to renew and nourish them.

To a single celled animal living in the water a better mode of locomotion than mere oozing along would prove exceedingly valuable. So in those called flagellates, mentioned in Chapter 2 (Serial Lesson 126), which are on the borderline between plants and animals, we find the cell secreting one or two hair-like lashes which carry them along swiftly by beating the water.

A certain amount of protection is afforded by a thin membrane enclosing a cell. Consequently, in a somewhat more developed form of protozoa, called the ciliated infusoria, such a membrane is secreted and the hair-like lashes which are used somewhat similar to oars are numerous. Also, as the containing membrane does not permit food to enter, there is an aperture in it, and in some forms, such as sedentary verticella, there are long lashes around this aperture that cause a whirlpool in the water and so bring the food down into the animal.

I have now mentioned members of three classes of protozoa. The phylum consists of four classes, each containing innumerable species. The fourth class developed more recently. Its members are parasitic, and unlike more ancient protozoa, they reproduce by means of spores. Each spore contains one or more minute germ. These germs and the animals they produce are the scourges of humanity, causing malaria, sleeping sickness, and a multitude of other dread diseases.

The ordinary protozoa and the cells of higher animals multiply by simple division. The particle of protoplasm contracts from two opposite sides, getting thinner and thinner in the middle until at last the connection is severed. In this process of division the nucleus of the cell always is divided, half of it going to form the nucleus of each new cell. When the two halves of the cell exist separately they gather food until both nucleus and its surrounding cytoplasm in each attain to normal size. The cells of the higher animals, including man, multiply in the same way as a primitive protozoan—by the mother cell dividing into two daughter cells—except that the cells of the protozoan go separate ways, and the cells of higher animals remain united.

Always, to explain the processes of higher animal life, we are compelled to return to the primitive protozoan, the first animal on earth; for in it we can perceive all the attributes and functions, in their simplest form, that we witness in the highest animal. But for the moment let us leave the protozoan and his single cell of living protoplasm and observe the formation of the first animal of more numerous cells.

Colonial Life

Some of the flagellates are considered the ancestors of brown algae, which are plants, and some are classified as protozoa. It is believed that a certain flagellate protozoan, on reproducing, instead of sending the daughter cells to some distant place, held them attached to the mother until there was a tiny plate-like colony of sixteen cells. These sixteen cells, each like a single celled animal, also each discharged all the vital functions. Yet because such an aggregation has certain advantages it was continued, and it came about whenever any one of the colonial cells reproduced by simply dividing, that the new cell went by itself, but dividing still further until it also produced a colony of sixteen cells. Such a sixteen celled colonial animal is the Gonium, and another whose colony tends to spherical form instead of being flat is the Pandorina. Both at the present day are common in fresh water.

Colonial life affording certain advantages, as time passed there came into being, in response to psychokinetic power, not merely sixteen celled colonies, but colonies composed of a great number of cells. With the enlargement of the colony it became increasingly difficult for every individual cell in it to perform all the functions of life. Already in certain protozoa, where the front differed in shape from the rear, when it divided to form two, each half was compelled to reproduce features that it did not possess. This ability of the soul thus had been acquired before the development of colonial organisms.

As pointed out in Chapter 2 (Serial Lesson 126), in addition to the drive to express itself more fully, the two great primitive desires of all life are the desire for food and the desire for reproduction. In a colonial organism both functions will be performed more successfully if certain members of the colony specialize in securing and assimilating food, and certain other members specialize in bringing into the world offspring. Such a division of labor for the first time, in so far as living forms are concerned, takes place in the Volvox. It is a hollow spherical colony of several thousand cells in a single external layer held together by gelatinous material and fine protoplasmic threads.

In the Volvox there are two kinds of cells. The one kind, called somatic cells, perform the functions of nutrition and locomotion. The other kind, called germ cells, perform the function of reproduction. The germ cells, through division, are able to form not only other germ cells, but also somatic cells, and thus when separated from their parent build up a new organism. This primitive division of labor also holds in the higher animals and in man. The ovum, which is a germ cell, always consists of a single cell. This divides into two daughter cells, and these into four, these into eight, sixteen, and finally into a cluster which arrange themselves into two strata forming a sack. From this stage, which has already progressed further than the Volvox, the forming organism passes through those stages of development parallel to still higher forms of life to be considered later; some of the cells secreting muscular tissue, some secreting the skeleton, some the nerve tissue, until the complete animal is present.

But in the Volvox there is still another division of labor, for the germ cells provide two kinds of sex cells, one male and one female. In the union of cells for the purpose of reproduction two things are essential; that the cells shall find each other, and that the resulting offspring shall be supplied with nourishment. To insure their union the male cells are very numerous, and as economy of material is advantageous, they are very small. In order that they may find the female cell they have the power of locomotion well developed. This locomotion, even in the higher animals, including man, is provided for by lashes similar to those of the flagellate protozoa. In fact, the sperm of higher animals has many points in common with the flagellates.

That the offspring may be provided with nutriment, the female germ cell specializes, not in movement, but in storing food. Consequently it is much larger than the male germ cell, as is markedly the case in the domestic fowl; for the yolk of a hen's egg, while still inside the hen and before fertilization sets up cell division, is but a single cell.

Contrary to popular conception, the sexual union of cells is not primarily to enable reproduction to take place, and originally had nothing to do with reproduction. Naturalists hold that its purpose is to enable the qualities of both parents to be inherited by the offspring, and Hermetic Initiates believe it further serves the purpose of revitalization.

During the sexual union of two protozoa there is an exchange of chromosomes (Chapter 4, Serial Lesson 167, Course 17, *Cosmic Alchemy*). When they separate the nucleus of each animal contains half of the chromosomes of the other and half of its own. This insures, then, when each cell divides in future, that the offspring shall, like the parents after fusion, contain the qualities of both. It also provides for another important attribute; for protozoa that from time to time enter into union continue to live and reproduce, or at least live and thus have opportunity for reproducing, while those that fail to do so die. Unless they meet with violent ends, protozoa that have the opportunity for union do not grow old and die. It might be well, therefore, for certain ascetic cults that herald from the housetops that union save for the rare purpose of reproduction is a crime, to pause and consider the biological fact, as stated by our best scientists, that the only animals on earth that are physically immortal can and do reproduce without union, but that union is absolutely essential to their physical immortality.

PHYLUM II, the Porifera: They embrace the sponges. The cells usually are arranged in the form of a hollow attached vase through the walls of which are many canals, or pores. This small vase, instead of being composed of a single layer of cells like the Volvox, is composed of three layers held together. Division of labor has here progressed further, with compensating advantages; for the cells of the inner layer have little hair-like lashes. In fact, they greatly resemble the flagellate protozoa. They lash the water, causing a current to flow through the canals, and also take in and digest the food thus brought to them. The middle layer of cells helps with digestion, and also secretes the hard framework. And even as some protozoa secrete lime, some silica, and some chitin, so there are sponges whose framework consists of each of these substances. The sponge of commerce is the chitin skeleton secreted by a whole colony of sponges. Such a colony held firmly together by a framework has the advantage of protection from enemies that would readily swallow and digest unattached individuals.

PHYLUM III, the Coelenterata: These embrace the hydroids, the jellyfishes, and the corals. The individuals commonly are called polyps. The body is a sack, in the center of which is another sack, an arrangement that facilitates digestion. Around this sack other membranes radiate. These radial members are usually tentacles, which assist in procuring food, and often assist in protection. The sea anemones, so common on rocky beaches near Los Angeles, are stationary polyps. Reef corals have the ability to secrete a skeleton of lime, which is securely fastened to the skeleton of their ancestors, making it difficult for their enemies to dislodge them. They are minute in size, but almost infinite in number. The reef, which is largely composed of their skeletons, rises at the rate of half an inch in ten years. The red, or pink, coral ruled by Venus, thought by the ancients to be a sure protection against evil influences when worn, is secreted by a coral called Corallina rubrum. The jellyfishes, which are colonial organisms, have developed the power of locomotion, which is an obvious advantage, and also in addition to feeding tentacles have others armed with stinging cells, such as are present in the Portuguese man of war, common in southern waters. In both digestion and defense the Coelenterata have made a distinct advance over the sponges.

PHYLUM IV, the Platyhelminthes: These embrace the flat worms. The flat worms are numerous on the land and in both fresh water and sea water, many kinds being parasitic. They are the first animals to have a right and left side and the first to have a front end which, although possessing no head, is carried forward. They have developed sense organs that enable them both to see and hear somewhat, which is a great advance over lower forms both in securing food and in escaping enemies.

PHYLUM V, the Nemathelminthes: These embrace the round worms. These worms are cylindrical in shape, and have a decided advantage over the flat worms in possessing a body cavity, which is a great aid in the digestion and assimilation of food. This valuable feature of an intestinal canal, however, often is lost when the species become a parasite.

PHYLUM VI, the Trochelminthes: These embrace the wheel worms, the rotifers or wheel animacules, of minute and various shapes. Some swim by means of hair-like bands which resemble revolving wheels. They are rather more complex in structure, and in this respect have made an advance over the animals so far mentioned. Yet their general features were not of sufficient value to be adopted by life as it developed further.

PHYLUM VII, the Molluscoidea: These embrace the Bryozoa, lamp shells and sea mosses. Such animals live in the water, the Bryozoa being a colonial form greatly resembling plants, common on our rocky beaches. They have various ingenious adaptations, and possess a well-developed digestive canal. Typical of this group is the lamp shell, abundant off the coast of Maine, and to be found near Los Angeles. The animal secretes a shell of two valves, which it opens and closes by muscular action. There is a mouth, and a groove bounded by little tentacles to guide the food to it. There is an esophagus and a stomach, and a stomach gland for performing digestion. The blood is colorless, and although there is no heart, contains corpuscles. It seems to be the precursor of the true mollusks, and has made an advance over lower forms in the matter of digestion and circulation.

PHYLUM VIII, the Echinodermata: These embrace the star fishes, the sea urchins, sea cucumbers, etc. They all live in the sea and are built on a symmetrical radiate plan such as gives the star fish its name. They have an outside skeleton, usually protected by numerous spines. They also have a great number of tube-like feet ending in suckers, by which they move, and in the case of the star fish by which they open the shells of their prey. There is a blood system, a nervous system, and also a water vascular system peculiar to themselves. This group is exceedingly well adapted to the environment in which it lives. They have not developed from any of the four groups of worms, but undoubtedly are superior modifications from Phylum III, the Coelenterata. Their chief advance over lower forms is the possession of a superior stomach and digestive system, and a superior circulatory system.

PHYLUM IX, the Annulata: These embrace the worms. There are a great many species of these, and they have made unusually important advances over any forms previously considered. Their bodies are elongated, and composed of ring-like divisions, each segment containing a separate and similar set of internal organs. There is also a blood system. Our common earth worm is a typical example. The sense organs of sight and hearing are more developed than in lower forms, and more important still, there is a nervous system having distinct ganglia, the first and largest ganglion being a part of the head. This, of course, foreshadows a brain, and is the most important advance over lower forms. The nerve chain is supported by a bundle of fibres which run along with it, and both are enclosed in a common sheath of connective tissue. The highly developed nervous system is advantageous in enabling a ready response to be made to environment, and some naturalists believe the sheathed nerve chain, which lies in relation to other organs as does the vertebrae in higher animals, which is also segmented, is the ancestor of the true vertebrate structure.

PHYLUM X, the Arthropoda: These embrace the crayfish group, the thousand legged worms, the spiders and the insects. They are animals possessing an elongated and transversely segmented body, with muscles attached to the inside of an external skeleton. This is quite the reverse from still higher animals, which have an internal skeleton about which the muscles are attached. Some of the body segments bear appendages, such as legs or wings, which are moved by muscles. The external skeleton is composed of chitin, a substance which certain protozoa also secrete. This hard outside skeleton prevents increase in size, hence growth occurs through shedding, or moulting, the chitin. There is a heart in most species, and a well developed circulatory system, as well as a suitable breathing apparatus. There is a mouth, intestinal canal, a brain, and a nervous system. The heart and brain are notable advances over lower life forms.

The crustaceans, such as the shrimp, crayfish and crab, which live in the water and breathe by means of gills are included in this group. It is probable as soon as plants moved out of the water in the Middle Ordovician period although the earliest so far found is the Psilophyton, a little plant about a foot high without leaves from the Devonian period which commenced about 350 million years ago that such animals as quickly followed this food supply found gills insufficient to supply them with the necessary oxygen for life. No doubt numerous experiments were tried through psychokinesis before this inner-plane power devised the expedient of having a system of tubes, called trachea, that with their microscopic branches permeate the whole body, air entering these tubes by external openings called spiracles. This system of breathing, because air reaches all organs and parts of the body, is in many respects superior to the lung breathing of vertebrate animals. It conduces to great activity, and is the system used by spiders and insects.

Insects have been unusually successful, 250,000 forms now being known with the tropics yet largely to be explored. They have made use of every available position in nature, have developed colors to protect themselves by concealment, have developed offensive weapons such as the sting of the hornet, and of still greater importance, because facilitating locomotion, is the common feature of wings. The most primitive insects, such as the spring tails, have no wings. Instead, at the end of the body are two elongated prongs which are bent under the abdomen and when pressed down form a lever by which the insect jumps. Such leaps, still further amplified in the flea and grasshopper, were undoubtedly steps leading to the development of true flight.

As but a single instance of the wonderful extrasensory intelligence displayed by insects, let us consider the wasp. The various digger wasps, in need of a food supply for their young, capture other insects with which they fill their burrows and on which they lay their eggs. Meat after being killed does not keep indefinitely, so these wasps, anticipating cold storage, devised a method by which their young might be provided with fresh meat as soon as hatched. They sting their prey in such a way as to reach the main nerve and paralyze the creature without killing it. The wasps of the genus Ammophilia have even gone beyond this and have arrived at the tool making stage of progress. After the burrow has been completed the female wasp fills it with paralyzed caterpillars and then packs earth over the opening, using a stone as a tamping iron (p. 21, *The Insect Book*, by Dr. Leland O. Howard) to pack this earth down. Later she visits the spot occasionally to see if all is well and to place disguising objects where they will conceal it. Such provision is taken for the young, even though in many cases the parents die before the young hatch out.

Instances of insect intelligence could be multiplied indefinitely. Ants, for example, keep slaves. They also keep the equivalent of cows, which they manage with great sagacity. These are aphides and in California the scale from which by stroking they get a sweet secretion. Some ants are excellent farmers, not only keeping plants not desired for their seeds from growing, but as in the case of the leaf-cutting ant, actually cultivating in prepared beds a species of fungus which is their sole food supply, and which must have been cultivated by them for an immense period of time, as it has never been found in the wild uncultivated state.

But of the various wonders of insect life none is more difficult to understand than the metamorphosis. Certainly the ability of the soul to live in and function through an inner-plane form after the dissolution of the physical is no more amazing. Primitive insects do not experience this change but hatch as miniature adults. More advanced insects show only a partial metamorphosis, the change from the larval stage being made by a series of moults that do not prevent feeding. But in the higher forms the insect hatches from the egg as a larva, which feeds voraciously and grows rapidly. Then comes the pupal stage in which there is no external activity, the insect being in a trance or comatose condition. While in this trance state the tissues are broken down and form a homogeneous fluid underneath the external skeleton of the insect. It thus precisely resembles the ectoplasm (Chapter 7, Serial Lesson 45, Course 1, *Laws of Occultism*) which emanates from a medium during materialization. This ectoplasm has been proved to be composed of organic substance drawn from the medium and sitters. At first it is plastic and without structure, but may materialize into a form of actual flesh and blood.

The caterpillar in its trance condition not only dissolves to a structureless plastic fluid, but this fluid is reconstructed along entirely different lines into a creature having almost no resemblance to its former self. Undoubtedly the ectoplasm from a human medium is organized by an intelligent agent employing the medium's astral body through which to exercise psychokinetic power. And it is equally certain that the soul of the insect acts through its astral form, in which its own and inherited experiences reside, in a similar manner.

PHYLUM XI, the Mollusca: These embrace the mollusks such as the clam, oyster, mussel, snail and octopus. In fact, it includes all the sea shells commonly found along the ocean beach as well as the slugs and snails found crawling in our gardens. The bodies are bilaterally symmetrical, unsegmented, and enclose a sack-like fold or mantle, which usually secretes the external skeleton, or shell. They are mostly able to crawl, swim, and burrow. They have a head, possessing a mouth and other appendages, with organs of special sense. Respiration is by means of gills. Quite interesting has been the discovery, through the study of embryology, that the young greatly resemble segmented worms, and in their growth show the steps by which the mollusks developed from such annulata. There is a good digestive system and a liver, which is an important advance. But as marking a still more important advance over previously mentioned forms is the development of a three chambered heart, and blood which in some species is red. This gives vigor of movement, which is a great advantage.

PHYLUM XII, the Chordata: These embrace the vertebrate animals; those possessing a backbone and those that show the presence of a primitive backbone at some stage of development. The main advance of this group, which includes all the higher animals, lies in the development of a second body cavity which houses a central nervous system; the spinal cord and brain.

Physical life has ascended from a single celled ancestor. One may observe closely the steps by which all present day plant structures are but the result of leaf modification. And a detailed study of the functions of present day animals is convincing that the animals now on earth developed from a single primitive cell. From the standpoint of religion this is an important finding, one replete with hope and assurance.

If man is a special creation, put here by an arbitrary Deity, there may be a hell to be dreaded, and a heaven which as usually described would be so monotonous that extinction would be preferable. But as all evidence goes to show that the soul actuated by the drive for sustenance, the drive for reproduction, and the drive for more ample expression, has developed a lowly single cell into higher animals and man, the possibilities of still further progress on the inner plane seem infinite.

Chapter 4

Fishes and Amphibians

Serial Lesson Number 128
Original Copyright, 1949
Elbert Benjamine
a.k.a. C. C. Zain

Approximate Geologic Time Scale
Recent Period to Cenozoic Era

Approximate Geologic Time Scale

Recent period has lasted to date	30,000 years
Recent period commenced	30,000 years ago
Pleistocene period lasted	970,000 years
Fourth glacial advance lasted	50,000 years
Fourth advance commenced	80,000 years ago
Third interval lasted	200,000 years
Third interval commenced	280,000 years ago
Third glacial advance lasted	50,000 years
Third advance commenced	330,000 years ago
Second interval lasted	425,000 years
Second interval commenced	755,000 years ago
Second glacial advance lasted	40,000 years
Second advance commenced	795,000 years ago
First interval lasted	175,000 years
First interval commenced	970,000 years ago
First advance lasted	30,000 years
First advance commenced	1,000,000 years ago
Pleistocene period commenced	1,000,000 years ago
Pliocene period lasted	6,000,000 years
Pliocene period commenced	7,000,000 years ago
Miocene period lasted	12,000,000 years
Miocene period commenced	19,000,000 years ago
Oligocene period lasted	16,000,000 years
Oligocene period commenced	35,000,000 years ago
Eocene period lasted	20,000,000 years
Eocene period commenced	55,000,000 years ago
Cenozoic era commenced	55,000,000 years ago

Approximate Geologic Time Scale
Cenozoic Era to Archaeozoic Era

Approximate Geologic Time Scale

Cenozoic era commenced	55,000,000 years ago
Mesozoic era lasted	135,000,000 years
Cretaceous period lasted	40,000,000 years
Cretaceous period commenced	95,000,000 years ago
Comanchean period lasted	25,000,000 years
Comanchean period commenced	120,000,000 years ago
Jurassic period lasted	35,000,000 years
Jurassic period commenced	155,000,000 years ago
Triassic period lasted	35,000,000 years
Triassic period commenced	190,000,000 years ago
Mesozoic era commenced	190,000,000 years ago
Paleozoic era lasted	360,000,000 years
Permian period lasted	25,000,000 years
Permian period commenced	215,000,000 years ago
Pennsylvanian period lasted	40,000,000 years
Pennsylvanian period commenced	255,000,000 years ago
Mississippian period lasted	45,000,000 years
Mississippian period commenced	300,000,000 years ago
Devonian period lasted	50,000,000 years
Devonian period commenced	350,000,000 years ago
Silurian period lasted	60,000,000 years
Silurian period commenced	410,000,000 years ago
Ordovician period lasted	70,000,000 years
Ordovician period commenced	480,000,000 years ago
Cambrian period lasted	70,000,000 years
Cambrian period commenced	550,000,000 years ago
Paleozoic era commenced	550,000,000 years ago
Proterozoic era lasted	450,000,000 years
Proterozoic era commenced	1,000,000,000 years ago
Archaeozoic era lasted	750,000,000 years
Archaeozoic era commenced	1,750,000,000 years ago

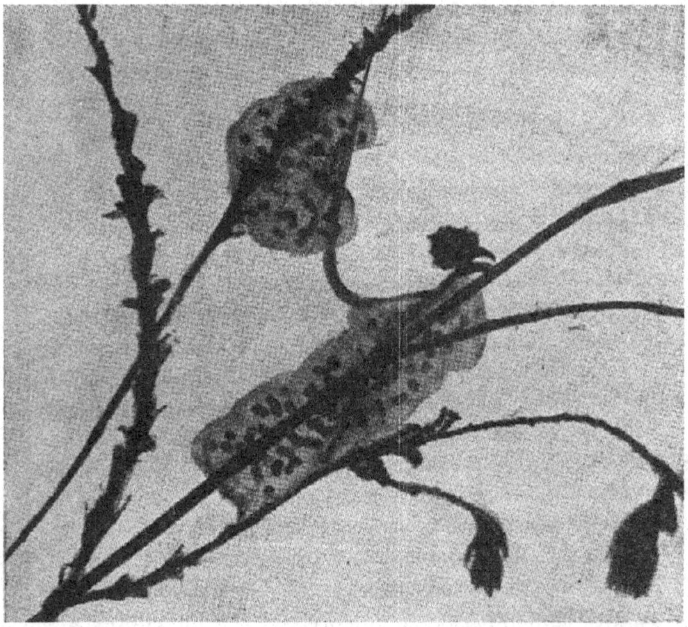

Tree Toad Eggs

All who have been in the country in spring or in the wet California winter, have heard the loud, "crack-it, crack-it, crack-it," of the tree toad (hyla). Not much larger than a man's thumb, by this call he attracts the female from the surrounding terrain to the water in which, as shown in the picture, her eggs must be laid.

Chapter 4

Fishes and Amphibians

MANY factors are involved in the evolution of life forms on earth. The soul attached to an organism through psychokinetic power has the desire to survive, the desire to reproduce, and the desire to express its potentialities through the organism. A certain life form may have so well adapted itself to its environment that most of its individuals are quite content. As has been pointed out, certain primitive bacteria, certain primitive plants, and certain primitive animals persist in vast numbers today in practically the same state of development their ancestors had reached a billion years ago.

Also, at the present day the majority of the people of the earth are making no strenuous effort to progress. Those who have adapted themselves to modern conditions sufficiently that they have an ample living and can raise their children in comfort and give them the conventional schooling, as a rule make no effort to advance farther. But there are a few individuals who never are content to reach a static stage. There are a few individuals with tremendous ambition to express themselves more fully, and who develop their natural aptitudes far beyond the average level. The important inventions, the great scientific discoveries, the new ideals, and the leadership of mankind in economic, political and humanitarian progress are the work of comparatively few individuals who, irrespective of economic security, are unwilling to drift with the tide.

From our study of their birth charts we know that these exceptional leaders were born, not from any different hereditary stock than others—for seldom is there more than one genius in the same family—but at a time when the inner-plane weather mapped unusual natural aptitudes. This means that their experiences in lower forms had given them a thought cell organization quite different than that of the average individual. This exceptional thought cell organization was mapped by the inner-plane weather coincident with their birth.

The other factor of significance is that subsequent inner-plane weather gave additional energy to the thought cells endowing them with these aptitudes, and thus gave them the psychokinetic power both to develop the natural aptitudes into special ability, and to attract into their lives the physical conditions that permitted them to use these special abilities in a manner that assisted in the progress of the human race.

Heredity

The nucleus of the reproductive cells carry filament-like chromosomes, which in turn are made up of bead-like strings of smaller bodies called chromomeres. The chromosomes, which in each species are of a definite number, are thus composed of many hundred bead-like bodies, the chromomeres, strung together in a very definite order. These chromomeres not only are the physical carriers of heredity, but in a given species of animal each bead in the string always governs some particular characteristic, and always occupies the same position in the string. The hereditary characteristics which are carried by these chromomeres are called genes.

In the union of germ cells of higher plants and higher animals, as in the union of two protozoa, when the cell divides after fusion each resulting cell contains chromosomes derived from both parents. The offspring thus obtains half its chromosomes from the mother and half of them from the father. One string of genes is supplied by the mother, and one string by the father, so that for every hereditary characteristic there are two beads, one of the pair furnished by each of the parents. Each member of any pair of genes, one of the pair being furnished by each parent, thus always has the same general function.

This doubling of the genes is nature's insurance against defectiveness; for commonly if one member of the pair is defective, the other sound gene has strength enough so that no defect appears in the offspring. The normal gene, because of its strength to impress its qualities on the offspring, is called dominant; and the defective gene, because it has much less power to influence the offspring, is called recessive. How from the combinations of dominant and recessive genes offspring inferior to both parents, offspring like superior parent, offspring like inferior parent, offspring superior to both parents, and genius may be born to the same parents is explained in Chapter 4, Serial Lesson 167, Course 17, *Cosmic Alchemy*.

These results are in conformity to Mendel's Law, which has a purely physical basis. But why certain chromosomes pair in one mating and other chromosomes, with dominant and defective genes different, pair in another mating is determined by the desires of the parents exerting a psychokinetic effect upon the germ cells as they thus unite.

But entirely apart from the chromomeres or genes that thus pair to continue hereditary traits, changes in the physical beads on the chromosome string often change the characteristics of the offspring markedly. These radical changes from the characteristics of any of their ancestors are called mutations. Such mutations are now induced by plant breeders for the purpose of developing new species.

A mutation exhibits some characteristic different from any in the ancestry of the individual displaying it, which breeds true. It once was thought that mutations were fortuitous changes issuing from heritage, but in recent years it has been found possible to increase their appearance by treating breeding stock with X rays, radium, heat and other environmental factors. As there are radioactive minerals in the earth in certain regions, these radiations are probably responsible for physical changes in the hereditary genes which gave rise to important mutations in the past.

In addition to mutations, now that experiments with colchicine have demonstrated that acquired characteristics can be inherited, it must be taken into consideration that if the parents were successful in changing either structure or habits sufficiently to meet new conditions, some of the offspring would be likely to inherit the same traits. Furthermore, not only heat and other external environmental conditions are capable of producing mutations, but in like degree so is the inner-plane weather.

The glands of internal secretion manufacture complex hormones from materials which they get from the blood and lymph. These hormones are released into the blood by nerve impulses. These nerve impulses which thus release the hormones may have their origin in the objective mind, in the desires and emotions of the unconscious mind without any objective awareness of them, or be stimulated by progressed aspects which constitute the chief inner-plane weather affecting an individual.

That a given hormone commonly is released in unusual abundance in human beings during the time a certain planet forms a progressed aspect has been thoroughly demonstrated. In fact, such endocrine reaction is one of the outstanding observed factors upon which Stellar Diagnosis is based. The other outstanding observed factor is the thousands of statistically analyzed birth charts and progressed aspects of people who have suffered from given diseases. And these hormones, which have been proved to be so susceptible to the influence of inner-plane weather, determine the size, shape and texture of the body, make for intelligence or its lack, give courage or cowardice, imbue with ambition or saturate with laziness, and in general force the given attitude toward life.

Thyroxin deficiency, for instance, in a child produces a condition in which it fails to grow. Except the skull, the bones and cartilage fail to develop, the abdomen projects, the skin is rough, dry and bloated, the temperature is low, the hair is thin, the nose is flattened, and the hands and feet are clumsy. If the thyroid gland is removed from a tadpole it grows to large size but never becomes a frog. On the other hand, a tadpole less than a day old may be transformed into a frog by feeding with thyroxine. It will retain its small size, but will have all the characteristics of a mature frog. Tadpoles and rats given the growth hormone of the front pituitary gland grow to huge size. When at puberty the gonad stimulating hormone of the front pituitary gland which responds to Pluto fails to increase its secretion, an adolescent boy becomes excessively fat with folds of tissue around his girdle and with prominent breasts, and his sexual organs remain infantile.

Without indicating the physiological changes that other endocrines inaugurate, when in excess or deficient, it will be apparent that astrological energies through thus stimulating or depressing the secretion of various hormones can exercise a powerful influence over structural changes of a life form. And it will be apparent that through influencing these secretions the soul has at its disposal agents through which important physiological changes intensely desired can be brought about either in itself or in its offspring. Any such marked progress of its own can be handed down through heredity to its progeny; and any change far beyond its power to bring about in its own physical structure, through mutation due to a variety of influences can be brought about in some of its posterity.

But it should not be concluded that each effort of this kind results in progress or the desired adaptation. Even as the diffused primal sense of touch became canalized and specialized, so extrasensory perception by which the unconscious mind of creatures apprehends things on the inner plane, to become serviceable must have exercise in effort and discrimination. Both it and psychokinesis in the lower forms of life are much like the faculties of a child soon after birth. The child sees something it wants. But as yet its experience has not taught it how to reach for the desired object or if it is at a distance how to walk or crawl to it. But it does make random movements, and when these random movements bring success it learns how to touch the things within reach which it wants, and later how to move to a desired object at a distance.

When it is stuck by a pin in its diaper it does not know how to remove the pin. But it thrashes about and yells, and as a result its attendant searches to find what the difficulty is and removes the pin.

And in much the same manner a life form dimly sensing changed environmental conditions and impelled by the desire to survive, uses its psychokinetic power in a haphazard manner. Most of the improvisations better to adapt itself or its offspring to the changed condition even as most of the experiments of a scientist in trying to make a new discovery or of an inventor trying to devise a new gadget result in failure. The record of the rocks reveals innumerable changes in form and function which have proved successful only for a limited period and then in the long run have proved disastrous. One after another life form has moved up a blind alley only to be stopped by a wall it could not surmount.

But a few of these attempts to adapt to new conditions have been successful. And these life forms have continued, and their progeny have made further useful adaptations, so that the line of progress unbroken still continues. It moved upward until it reached the form of man, a miniature copy of the universe. Then it continued upward, not through further development of form, but through progress in knowledge.

Cooperation Between Plants and Animals

But even forms of life lower than man have not confined their efforts to survive, to reproduce, and to express more fully, to changing form and function. Even as man makes use of other life forms so plants often make use of animals and animals often make use of plants. Plants in general found it of great advantage to reproduce by means of seeds instead of spores; and they found it a further marked advantage when the male element from one flower could find its way to the female element of another flower and thus prevent inbreeding. Consequently, the plants, using such inner-plane intelligence as they possessed, sought how this vitally important cross fertilization might be accomplished.

The first expedient was to use the wind and water to carry the fine grains of pollen, which contain the male sexual element, to adjoining, or distant flowers. To insure pollination in this hit or miss manner, however, required great quantities of pollen; for much is sure to be wasted. Therefore, as insects were in the habit of visiting the plants, which they used for food, the expedient gradually was adopted of using these insects to carry the pollen from one flower to another.

The problem then arose of making sure that insects would come to the flower. This was solved by secreting a substance, such as the nectar found in most flowers, that would serve the insect especially well for food. Next the problem arose of attracting the insect from a distance. This was accomplished by coloring the flower, or by giving it a pleasing odor, that might be noticed by the insect at a distance. The color of a flower, and its odor (unless this is offensive and used for protection) have been developed for a single purpose and no other; to act as advertisements that a banquet awaits the particular insect best suited to carry out its cross fertilization. These advertising banners have been subject to a special evolution. Thus the most primitive color for a flower, aside from the original green, is yellow. Later, plants developed red blossoms, and finally, as the very latest color scheme, and one that is recognized farther than the others, the blue and violet colors have been evolved. In some way, although not exactly in the way we discern colors, but perhaps by perceiving degrees of brilliancy, insects are able to distinguish the blossoms of their favorite flowers at long distances.

With the problem of attracting insects solved, the next step was to make absolutely certain that the insect securing the repast of nectar should pay for its meal by being dusted with pollen. Gradually an infinite number of cunning devices, in the form of differently shaped corollas, were evolved. Flowers that originally had numerous petals, and numerous pistils and stamens, securing fertilization in some few by sheer numbers, reduced the number of parts in favor of some shape that would permit the insect to enter, but which also insured that it carried away pollen. All the innumerable forms of our bright colored flowers tubes with little landing platforms, hoods, sheaths, and what not have been evolved in response to the plant's intense desire to compel the insect upon which it depends for cross pollination to carry pollen from the male portion of one flower to the female portion of another flower.

All those plants, then, that have small greenish inconspicuous flowers, like the grasses, depend upon the wind for pollination. The bright ones invariably depend upon insects or birds. Nevertheless, some that in the past depended on insects have now devised an unusually fine, light pollen which is produced in enormous quantities. At the time of bloom the air for miles around is full of this minute pollen dust. These plants, of which the goldenrod and ragweed are typical examples, are finding this new method even superior to depending on insects. They are, therefore, abandoning the use of insects and returning to the use of wind, being now in the state of transition.

The problem of dusting an insect with pollen solved, the next thing was to make sure that the pollen would be deposited not on the female part of the same flower, but on the female part of some other flower of the same species. This objective is attained in many unique ways. For instance, the little filaments, or stalks, of the stamens of the cornflower, when touched, contract and draw instantly down over the stigma, or female part, protecting it; yet at the same time exposing the pollen on the anther to the insect which has touched it. The diplacus, or monkey flower, common on our California hills, has a stigma of two flat lobes that snap together tightly when touched. The stigma is placed so prominently that an insect visiting the flower is sure to touch it when alighting. It snaps shut on the pollen dusted from the startled insect, and the insect then gets covered with the pollen from the anthers of this flower with no danger that any of it will find its way into its own stigma.

To insure that they shall not be fertilized by their own pollen, the pollen of many flowers ripens only at a time, either before or after, when the stigma is not yet able to receive pollen. The larkspur has still another device. It bends down certain of its stamens on different days, so that if the bees that visit it do not on one day visit and fertilize other flowers with its pollen, those that visit it several days after this get still another load of fresh pollen. The larkspur, too, belongs to the buttercup family, and practically all of our wildflowers and garden flowers are thought to be descendants of a primitive buttercup. This early buttercup had numerous petals, numerous pistils, and numerous stamens; which were gradually sacrificed in the interest of greater efficiency. The modifications, in each instance, were toward securing pollination through the aid of some special insect or bird.

The hummingbird sage, common about Los Angeles, has developed such a long tubed corolla that few insects can reach the nectar in the bottom. It is a favorite flower of the hummingbirds, however, which its crimson blossoms attract from long distances; and it depends chiefly upon these for carrying its pollen. Other flowers depend upon bees. Here another problem arises; for if a bee visits one kind of flower and then another kind of flower, the pollen so carried will not fertilize. The pollen from white clover, for instance, will not fertilize the buckwheat flower which next may be visited. In this case it is the bee that has learned something; for as plants have progressed, so have the insects that live upon them progressed in parallel manner. Bees are absolutely dependent upon nectar and pollen for food. It is to their advantage that flowers shall be pollinated properly, thus providing for new plants to blossom the next year. And it has been definitely determined that bees do not indiscriminately visit different kinds of flowers. During the work of a morning a bee will confine its attention to one kind of flower. It does this even though it visits different colored flowers of the same species. On the next day it may turn its attention to a different species of flower, but it is too good a gardener to mix the pollen of a daisy with that of a dandelion.

Flowers that open by day depend upon day flying insects. Those that open at night depend upon insects that fly by night. Those with the nectar in short tubes depend upon bees and small insects. Those with long tubes depend upon butterflies and insects with long probosces. The red clover, for instance, depends entirely upon the bumble bee. Efforts to raise red clover in New Zealand were a failure until some naturalist suggested importing bumble bees. Plenty of bumble bees insured proper pollination and made the venture a success.

One might write on indefinitely of the manner in which flowering plants and insects have helped each other solve the problem of life and progress. One might write on indefinitely of the shrewdness of insects, such as the common harvesting ant of South Europe, which collects the seeds of clover like plants, lets them sprout until they burst, then exposes them to the sun to prevent further germination, after which it carries them under ground. Still later it chews them into dough and makes them into little biscuits which it bakes in the sun. These it then stores for winter use. Such wonderful habits, which in many cases parallel the efforts of humanity after reaching some degree of culture, are the outcome of extrasensory perception combined with experiences stored as knowledge within the soul, directing psychokinesis in response to the desire to live and express more fully.

Fishes Were the First Truly Vertebrate Animals

Now moving back in time before there were any land plants and before there were any insects, in the Archaeozoic era there were only single celled animals and single celled plants. Then came the Proterozoic era, which commenced about a billion years ago, and during which invertebrate animals developed innumerable forms which dominated the world. Following this was the Cambrian period of the Paleozoic era which commenced about 550 million years ago, during which the trilobites became the dominant life form. These are segmented animals belonging to the phylum Articulata, having for ancestors the segmented worms. They are primitive crustaceans, and other crustaceans like the lobster, crayfish and shrimp developed from them. The trilobites are the transitional form between the segmented worms and the insects; for after land plants developed the descendants of the trilobites took to the land to get a food supply and gradually became insects as we know them.

But before the trilobites, undoubtedly there were segmented worms. Let us now, therefore, visualize a world covered with shallow seas and lakes, crowded with innumerable kinds of invertebrate life. Then let us imagine the condition, as actually transpired, when numerous land areas the world over commenced to rise. Instead of placid lakes, large areas tilted up to form highlands and were drained by swift flowing rivers. At the same time the lake and sea expanses, already crowded, were greatly reduced in area and forms of life that had found a living there were sorely pressed for food. Many such forms finding the competition too keen died out and became extinct. But other more progressive kinds tried to adapt themselves to the new condition by finding a habitat in the rivers.

Rivers have a persistent and rather rapid flow of water in a fixed direction. To be able to live in a river, and not to be washed down it and out to sea, an animal must either have some means of clinging to the bottom or some means of locomotion sufficiently effective to overcome the current of the stream. Except for certain minor instances there are only three large groups of animals that have solved this problem. Some of the mollusks, like the mussel, are able to crawl along the bottom through firm contact with it. Certain crustaceans, like the crayfish, can crawl along the river bottom by means of many sharp claws that they hook into the river bed. The fish solve the problem by a mode of propulsion through the water.

When the segmented worms took to a life in the river to escape their numerous enemies and to find food, they found that the position enabling them best to meet the current is to keep the head directly upstream. They found also, by degrees, that a rhythmical undulation, similar to that of grass growing in the stream, is the movement best suited to overcome the momentum of the current. A fish moves by alternate rhythmical contractions of the side muscles, so that the pressure of the fish's body is brought to bear, first on one side and then on the other side, against the water of the incurved section. Such motion is not possible to most invertebrates of the sea; for usually they have compact or rotund bodies that make them sluggish. But the segmented worms had a suitable linear form, and already had a bundle of fibres running lengthwise with a nerve chain enclosed in a sheath. This was the commencement of a lengthwise supporting tissue that would prevent the shortening of the body due to the pressure of the water against the head.

There developed, therefore, such creatures as the enteropneusts, which are vertebrate-like worms. They have numerous gill slits opening from the pharynx to the back surface of the body, and a body cavity similar to true vertebrates. They live at present off the coasts and eat their way through the sandy mud to get the small organisms living in it. A somewhat further development from the segmented worms is shown in the sea squirt. It begins life as a free-swimming larva, like a small tadpole, with a brain and spinal cord, a notochord, or primitive vertebra, a brain, eye, and a heart. It fastens itself to a small shell or stone, and then degenerates rapidly as it reaches the adult state. Next above these come the lancelets, such as Amphioxus, that are found in most seas. They have no skull, no jaws, no limbs, no brain, no heart, and no eye; but they do have a spinal cord, a notochord, and gill clefts. They are translucent spindle-shaped creatures about two inches long that are believed to have a worm ancestry, but which have developed the ability to swim with some speed.

A still more advanced transition type between segmented worms and true fishes are the round mouths, such as the lampreys and hags. The lamprey looks something like an eel, but has no jaws, no limbs, and no scales. It does have, however, a gristly skeleton, something of a skull with horn-like teeth, and a number of gill pockets. Some live in fresh water, and some live in the sea, ascending the rivers to spawn. The hag is another eel-like creature, one, the Bdellostoma, living off the California coast.

The first truly vertebrate animals were the fishes. These vertebrates are the group embraced in phylum XII, the Chordata. They have several distinct characteristics that separate them from all other forms of life, and as all the higher animals, including man, belong to this group, it may be well to mention these characteristics. In the first place, in the earlier forms there is a notochord, or primitive backbone, running lengthwise of the body, serving to stiffen it, and thus prevent the shortening of the body which otherwise would take place when the muscles are contracted. This notochord is composed of membranous connective tissue. In more advanced forms it is formed of cartilage. And in those forms still higher it becomes a bony vertebra column.

A second characteristic of all vertebrate animals is the development of gill slits through the walls of the throat cavity. We have already seen that these gill slits are present in the vertebrate-like worms. They are obvious in the true fishes, the gills being surfaces of considerable area where the blood is exposed to the oxygen contained in the water and respiration accomplished. In the mammals, including man, several pairs of gill slits are always well defined in the embryo, but as the form develops they are modified until a single pair are left, and these are no longer used for respiration, but form the eustachian tubes which connect the middle ear with the throat cavity and thus equalize the air pressure on either side of the ear drum.

All vertebrates also have a spinal cord, are usually segmented, and when paired limbs are present there are never more than four.

Most scientists believe the first fishes probably were not bony, but were gristly, with a mouth on the front side, like the sharks of today. Dr. W. K. Gregory, however, has advanced a theory according to which the Ostracodermi, which are fishes with an armor of large bony plates around the head, were the first vertebrates. Up to the present time no fossils of the earliest true fishes have been found. The earliest fossil fishes had traveled a long way along the road of vertebrate progress. The mentioned Ostracodermi left their fossils in the Middle Ordovician rocks, formed about 445 million years ago near Canon City, Colorado, in the Big Horn Mountains of Wyoming, and in the Black Hills of South Dakota. The Ordovician period is the period immediately following the Cambrian period in which the trilobites dominated the earth. Before the Cambrian, in the rocks of the Proterozoic era, which commenced about one billion years ago, there are traces of one celled marine animals, the radiolarians, with shells of flint that could be preserved. And there are also traces of worms that burrowed in the mud; for these burrows are sometimes preserved as fossils. But with the coming of Cambrian times the seas and shores the world over began to swarm with sponges, jellyfishes, crustaceans, worms, lamp shells and mollusks.

Certain of these mollusks, the cuttlefishes, dominated the sea during the Ordovician period. They were fierce predatory creatures, even as is the octopus at this day, but their place as masters of the sea was disputed by the true fishes that developed at this time, and they finally had to yield to them. These first fishes were fresh water fishes, and it is thought that the habit of many marine fishes today, such as the shad, sturgeon, and salmon, of leaving the sea and ascending rivers to spawn is the following of a custom established early in fish history. After developing locomotion, and the typical fish form, the fishes were better adapted not only to a life in swift moving streams, where they had their origin, but to water life in general. As a consequence they sought out every available nook of lake, sea and river, and so modified their structure as to make them specially suited to survive in the chosen habitat.

Intelligent Adaptations

Space will not permit of even a superficial enumeration of the various wonderful adaptations accomplished by fishes. Many are quite unique, such as the one called "the angler" which has a fishing rod and tempting bait which it dangles in front of a cavernous mouth lined with teeth that are hinged at the base so as to bend backwards, permitting other fish to enter but quite preventing their exit. The eggs of the sea horse are placed in a skin pocket bringing to mind the skin pocket of the kangaroo where they are sheltered until developed. Some fishes also make nests, anticipating the birds in this respect. The stickleback, for instance, makes an elaborate nest of leaves and stems of water plants which he sticks together with glue-like threads which are secreted at this time by his kidneys. This nest has two doors, and by coaxing and by using a certain amount of force, he persuades one female after another to pass into the nest by one door and out of it by the other, depositing her eggs in the nest as she goes. After this he sets himself to guard the nest, and drives away all other fishes that approach. After the young are hatched he is kept very busy herding the little ones together and keeping them out of danger until they are old enough to shift for themselves. This he does with the utmost diligence and solicitude.

Another male fish, the scientific name of which is Semotilus Atromaculatus, takes stones from the bottom of a stream, gripping them in his mouth, and builds them into a dam. Below the dam he builds an egg depository of stones so formed that the eggs when deposited by the female are held in the spaces between the stones, thus protected from other fish, and kept from being washed out by the dam just above them. Innumerable other examples of extrasensory intelligence as exhibited by fishes might be cited, but these two no doubt will suffice to show that the intense desire of the soul, even the soul of a fish, at times is able to find a means to the sought for end.

I have mentioned that the earlier fishes had a gristly structure and that those more developed had provided themselves with a bony skeleton. But there is yet another group of fishes that now needs to be mentioned. These are the Dipnoi, or double breathers, represented by the bony pike in the United States, and by the lung fishes of Africa, Australia, and South America. These lung fishes live today in regions where the lakes and ponds at one season of the year dry up. No doubt, in answer to the desire to survive in such an environment, fish were developed with the air bladder connecting with the gullet. In other fishes the air bladder serves as a means by which the fish rises or descends in the water, expanding the sack to rise and compressing it to sink. But in the lung fishes, when the pool in which they are living dries up, it is used as an accessory apparatus by which the blood is given oxygen from the air. These lung fishes can successfully weather long periods of drought.

There is also a fish, the "climbing perch," which abounds in fresh water throughout the Malay countries, Ceylon, India and Burma, that has the habit of leaving the water and traveling across the land, even over high hills and broad prairies, not infrequently climbing up trees on the way, to other water. This fish carries water in chambers of its head for the purpose of breathing. There is also a climbing catfish in the upper Andes of South America.

In an environment such as the present African mud fish live in, where the dry season lasts nearly half the year, a great premium is placed upon the ability to breathe air, and also to move about on land; for the water completely dries up. The persistent desire to survive and express more fully brought psychokinesis into play and undoubtedly developed the first amphibian, or land vertebrate, from the lung fishes.

In the Ordovician period the first fossil fish are found. In the next period, the Silurian, are found primitive scorpions, some of immense size; and it is quite certain that some of these took to life on the land; the segmented worms probably accomplishing this at an even earlier date. The oldest fossil amphibian is the footprint of Thinopus, found in the period following the Silurian, the Devonian period, which commenced about 350 million years ago. During this period, also, for the first time, flowering plants became established.

The next period after the appearance of the amphibians is the Carboniferous period, embracing the Mississippian and Pennsylvanian, during which the great coal marshes were laid down. Unlike the Devonian period, which was marked by aridity, there was a mild, moist climate that encouraged luxurious vegetable growth on low, swampy ground. This vegetation was mostly club mosses and horsetails that grew to immense size. Their spores and other debris is the source of the present day coal supply. The first fossil insects are found in the same period and undoubtedly the land swarmed with them. They provided certain cross fertilization for the flowering plants that had now become established, and they became a food supply for the amphibians which followed them over the land. It was no doubt at that time that the flowers first began to gain their colors and attract insect visitors. There were also land snails at this time. But perhaps of chief importance were the amphibians, some of which grew as large as a donkey.

The paired fins of the lung fishes, in response to the desire to facilitate eating, were gradually or perhaps more abruptly developed into limbs with fingers and toes, by which things might be grasped and food placed in the mouth. There was the development of an ear of three chambers, a movable tongue, true lungs, a drum to the ear and lids to the eyes, none of which a fish has. Furthermore, for the first time there was developed a voice. At first the voice served as a sex call, as it does today with our toads, hylas (tree toads), and frogs. The piping and croaking of these amphibians, so noticeable in the spring of the year, are love calls. As higher forms of life developed the voices came to be used to express a call for help, to convey the notion of danger, and to express other emotions.

The amphibians, represented by our frogs, newts and salamanders, are air breathers in the adult stage. They must return to the water to lay their eggs. The young are hatched in the water, and pass through a fish-like period of infancy, breathing by means of gills. They thus, in the early stage of their lives, recapitulate their development from fish ancestors. Every schoolboy is familiar with the tadpole that lives in the water and later absorbs its gills and tail, gains four legs, and transforms into a frog or toad able to travel and live on the land.

The Record of the Rocks

I have already been compelled to use geological eras and periods to designate the time in the earth's history when certain forms of plants and animals first developed. It will be well, therefore, before going further, to explain how we know the comparative ages of these periods, and how we know that certain life forms first occurred at stages of the earth's history corresponding to them in time.

We have all watched, during a rain, the tiny rivulets running down a hillside, cutting little gutters in the soil and carrying sand from the hill to a creek, thence into the river, and finally to the sea. This process is going on yearly, and at times great rivers go on a rampage and in a short while cut down their banks and carry great quantities of mud and sand into the ocean. Creatures that have died during such a flood often are carried into the sea and buried beneath the sand and mud. As time flows on they are buried deeper and deeper, until a great quantity of material lies above them.

This material is compressed and hardened by the accumulating weight above it until the sand becomes sandstone and the mud becomes mudstone or shale. If the creature thus buried has a bony skeleton, or other hard parts, and is buried in a deposit of lime forming near the shore, or in mud so that air and water cannot reach it, these hard parts are preserved in the forming rock as fossils. Sometimes, also, insects are caught in the resin exuding from trees, and encased in it. These trees then may be torn from the banks of streams by a raging torrent and buried in the mud. The mud then becomes stone, preserving the tree in it, and the resin turns to amber which encloses and effectually preserves even delicate insects. Volcanic dust occasionally overwhelms insects flying above shallow pools of water, bearing them down into the water and covering them with a layer of powdered stone that solidifies, encasing them in a hard shroud that effectively preserves them. They then become fossil insects.

The process of erosion, and deposition of sand and mud in the sea, is not uniform, but periodic. At more or less regular intervals great quantities of silt are deposited, and at other periods there is very little. Thus, the sand and mud is laid down in layers; for deposits of one period often harden somewhat before the deposit of the next period is put down. The succession of layers is easy to determine, as one will learn by watching the fan-like deposits of a hillside rivulet when it fails to reach a larger stream and must drop its load. Every rain increases the thickness of the deposit by one layer, and this new layer is always laid down on top of the layer previously formed by the preceding rain. The oldest layer of sand and mud always is on the bottom and the newest always on top. Thus it is also with the mud and sand laid down in the sea. Those layers on the bottom are the oldest, those next above these are next oldest, and so on, until the top layer is reached, which is the newest.

If fossil remains are found in the oldest strata, those creatures were buried at the time the oldest layer was laid down, and must have lived at that time. If other creatures in a fossil state are found several layers up from the bottom, these creatures were buried at a later date than those buried in the bottom layer, and consequently lived at a later date. Those found buried in the top layer were buried at the very latest period during which the deposit was formed, and thus must have lived at the latest time during which the deposit was formed.

All sandstones and shales and other sedimentary rocks were formed by being laid down as fine material in water and later solidified by pressure. They form definite layers, one above the other, and even though at a later date these layers are tilted up by the elevation of a portion of the area they cover into hills, the order in which they were laid down is not difficult to ascertain if cuts have been made through them. Furthermore, the layers laid down at different times vary, not only in the kind of life found fossilized in them, but also in their structure and mineral composition, so that it becomes possible for one skilled in such work to say with great precision just which of these layers of sedimentary rocks are oldest, which are next formed, etc.

The various layers of rock that were formed by sedimentary deposits in North America have been carefully measured. Their combined thickness is estimated by some authorities to be as much as 67 miles, but an average of the estimates of the various authorities gives their total thickness as 53 miles. Most experts believe this to be very close to the correct figure.

It should not be thought, of course, that in any one spot the sedimentary rocks are 53 miles thick. This is the thickness of all sedimentary rocks in all localities that have been deposited at different times. In Ontario, Canada, the sedimentary rocks are 18 miles thick, but throughout one third of North America whatever sedimentation there was has been eroded away, leaving igneous, or crystalline rocks at the surface. And it is known that at the present time it takes 8,600 years to denude North America one foot. The igneous rocks left bare by erosion and the tops of the Sierras in California are thus bare, such gold as their previously overlying sedimentary rocks contained having been carried down the streams to be found in part by those who stampeded west in 1849 have cooled to their present state from a molten or plastic state. Over the balance of North America the sedimentary rocks are from one mile to twenty miles thick; perhaps but one eighth of the area along the troughs adjoining such mountains as the Appalachian and Rocky Mountains attaining the greater thickness.

These layers of rock have certain structural characteristics by which they can be recognized. The order in which they were laid down has been determined by careful study of the relative positions of their layers. For convenience in speaking of them the whole system of rocks has been divided into five great groups classified according to age. The great groups are called eras, and are each divided into several periods.

The rocks of these five geological eras have been carefully measured. The earliest era, the Archaeozoic, commencing about 1,750 million years ago and lasting about 750 million years, consists of nine miles of limestones and nine miles of mud and sand stones. The next era, the Proterozoic (sometimes called the Algonkian), commencing about one billion years ago and lasting about 450 million years, consists of one mile of limestones and 13 miles of coarse mud and sandstones. The rocks above these are the Paleozoic, which commenced about 550 million years ago and lasted about 360 million years. It consists of 3.4 miles of limestones and 4.6 miles of mud and sandstones. Next higher is the Mesozoic, which commencing about 190 million years ago lasted about 135 million years, and embraces 1.25 miles of limestones and 6.25 miles of coarse mud and sand stones. The latest era, the Cenozoic, which commenced about 55 million years ago and lasted about 55 million years, consists of five miles of coarse mud and sandstones. The other half mile to make up the 53 miles of thickness consists of small and undetermined formations.

Chapter 5

Reptiles and Birds

Serial Lesson Number 129
Original Copyright, 1949
Elbert Benjamine
a.k.a. C. C. Zain

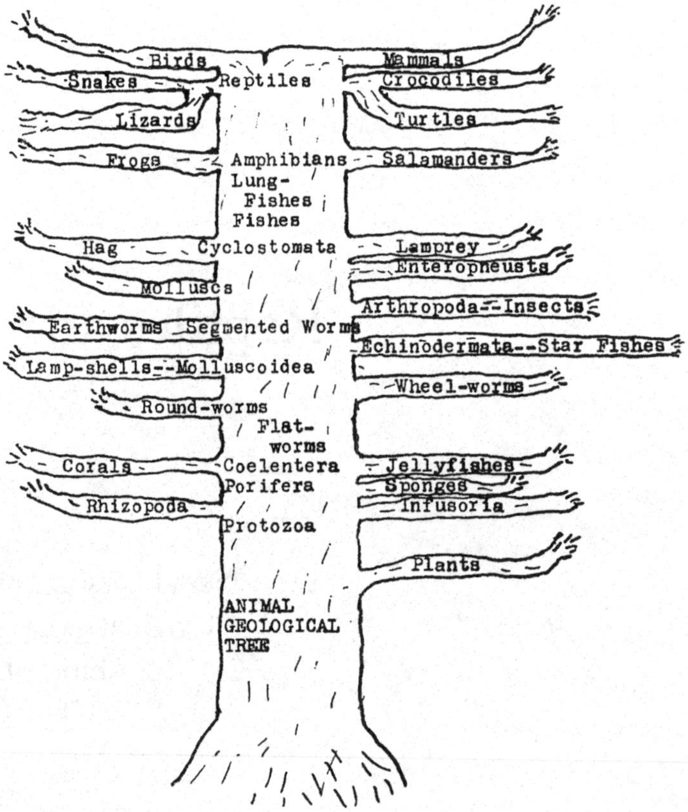

Animal Geological Tree

Recorded in the Rocks

Years Ago	Era or Period	The Record
1,750,000,000	Archaeozoic era	First record of bacteria.
1,350,000,000	Archaeozoic era	First single-celled plants.
1,250,000,000	Archaeozoic era	First animals; protozoa.
1,000,000,000	Proterozoic era	First record of marine algae, worms, sponges, crustaceans.
550,000,000	Cambrian period	Record of main invertebrates.
440,000,000	Middle Ordovician	First vertebrates; fishes.
440,000,000	Middle Ordovician	First land plants.
410,000,000	Silurian period	First air breathers; scorpions.
410,000,000	Silurian period	First lung fishes.
350,000,000	Devonian period	First record of amphibians.
350,000,000	Devonian period	First record of flowering plants.
300,000,000	Mississippian	Abundant sharks.
255,000,000	Pennsylvanian	First record of insects.
255,000,000	Pennsylvanian	First indication of reptiles.
215,000,000	Permian period	First metamorphosed insects.
215,000,000	Permian period	Indisputable reptile records.
215,000,000	Permian period	Indisputable records of modern ferns and conifer trees.
190,000,000	Triassic period	First record of dinosaurs.
170,000,000	Upper Triassic	Record of flying reptiles.
170,000,000	Upper Triassic	First record of mammals.
135,000,000	Upper Jurassic	First record of birds.
120,000,000	Comanchean period	Rise of flowering plants; extinction monster dinosaurs.
90,000,000	Cretaceous period	Record of first placental mammals; extinction all dinosaurs.

Recorded in the Rocks

Restoration of Archaeopteryx

Restoration of Archaeopteryx, the earliest fossil bird. It lived about 135 million years ago in the Upper Jurassic period.

Chapter 5

Reptiles and Birds

CHROMOSOMES, as the physical carriers of heredity, have already been considered. And before their discovery two other laws of heredity had been found that need mention. One is Galton's Law, purporting to show that ½ the qualities of an individual are derived from its parents, ¼ from its grandparents, ⅛ from its great grandparents, etc. The other is Mendel's Law, indicating how inherited characters are distributed, and that the fundamental cause of this distribution lies in the germ plasm. This law was worked out by Mendel on peas, and later by others on rabbits, guinea pigs, fowls, beetles, and silkworms.

It shows that inheritance is through unit characters—which has now been verified by a study of genes (Chapter 4, Serial Lesson 167, Course 17, *Cosmic Alchemy*)—and that while it cannot be predicted what proportion of white and black will appear in succeeding generations from mating white and black individuals, yet once the proportion is known, it may be used with certainty as the basis of future production. Thus if in mating a white and a black ¼ of the offspring are found to be white and ¾ black, it may confidently be expected that again in mating white and black of the same species the same proportion of white and black will appear in the offspring.

But in order for a new species to be formed it is necessary that some of the offspring show marked tendencies that neither of their parents had. The mechanics of these sudden appearances, called mutations, has been explained in previous chapters. It is enough here to point out that whatever change of structure or functions such a mutation possesses is handed down to its progeny through the indicated laws of heredity.

Natural selection, which was given detailed study by Darwin, is based upon the principle that those creatures that are somewhat better fitted to live in a certain environment survive, and those less fitted perish. Furthermore, for the same reason, an organ or a function that gives an individual an advantage, and therefore makes him more able to survive than his fellows who do not possess the organ or function, is preserved and passed on to his offspring.

The offspring, then, by virtue of their superior adaptability, live and propagate, while less fortunate forms of life that compete with them for food and life become extinct.

Due to the prolific tendencies of nearly all life forms, there is available space and food on earth for only a very small proportion of those that are born. As a consequence of the limited area and limited food supply, there is a bitter competition between the limitless forms for possession of this space and food. At the date of this writing (1949), for instance, Milton S. Eisenhower, president of Kansas State College, states that only 11 percent of the world's land can be used for food production—a total of 4,000,000,000 acres. As there are about two-and-a-quarter-billion people now living, this means there is less than two acres available for each. Yet it takes not less than two-and-a-half acres to raise enough food for one person. Eisenhower makes the flat statement that today more than two thirds of the people of the world do not get enough to eat.

But considering physical life as a whole, the competition for space and food is in three different directions: First there is the struggle between different species. Of two species deriving their food supply from the same source, or occupying the same territory, the one most suited to that particular environment will survive and leave offspring, and the other will perish.

This principle is well exemplified in the struggle between sheep and cattle in the western United States. Cattle once throve in vast number on the broad western ranges. There was plenty of food, and they multiplied rapidly. Then came the sheep. These could bite the grass clean to the roots, as well as live on brush and other unpalatable fare. They also left an odor where they passed that is disgusting to other herbivorous animals. Thus the cattle were forced out, for they cannot live where sheep thrive.

The owners of the cattle and the owners of the sheep then took up the struggle. The cattlemen, for a time, were greater in number and were skilled in the use of firearms. They realized that cattle could not compete with sheep, and made the issue in several sections of the West a war between cattle men and sheep men. This struggle later was carried to the courts, and finally to the legislature. Sometimes the victory has been with one and sometimes with the other, but the struggle is due to the fact that sheep have developed more ability than cattle to survive when there is keen competition for food.

Next, there is keen competition between the individuals of a single species. This is inevitable, because the rapid rate of multiplication, if unchecked, soon fills all available space and becomes greater than the food supply. Our two worst pests—because they carry deadly bacteria and deadly protozoa to the human race—the housefly and the brown rat, for instance, multiply at an enormous rate. It has been calculated that the offspring of a single pair of houseflies, if they all lived and reproduced and these lived and reproduced, would in a season be so numerous that their mass would be larger than the whole earth. The offspring of a single pair of brown rats, if all reproduced and none died, in a single year would number one million. It will be seen that at such a rate of multiplication there soon would be no food left, even though flies and rats had no other competing species to reduce their food supply.

We also see this competition between members of the same species in the circle of our human acquaintances. There is a struggle for money, for honors, and for special privileges. Some have millions of dollars and some have none, nor can find opportunity to earn their daily bread. We call ours a civilization, but surely it is not that so long as it deprives a single willing person of the opportunity to work for enough food to live. War, too, a barbarous ancient institution, is the direct result of this competition between individuals who have grouped together for mutual advantage. But the problem of the individual's food supply and the problem of war cannot be solved by any amount of sentimentality. They can only be solved by a discriminating use of carefully collected biological and sociological facts.

The third form of competition, about which considerable will be said later, is the struggle of both individuals and species with the forces of nature. Changes of temperature, in the amount of moisture, storms, and changes in elevation and amount of land areas, all require fresh adjustments of organisms living in a region where such changes take place. The struggle to meet the conditions imposed by environment results in a constant shifting of living things, only those surviving that are peculiarly adapted to the conditions that obtain.

Not only was natural selection given detailed study by Darwin, but Lamark gave detailed study to adaptation, and de Vries brought to scientific attention the significance of mutations in making successful adaptations which would be retained, through inheritance, by subsequent generations.

Before the significance of mutations was understood it was held that those individuals who developed characters that were slightly advantageous in competition with their associates lived and produced offspring, and that those less well adapted died before reproducing. But there are innumerable developments in life forms that would produce very slight advantage until the new and highly complex structure was fully present. The simple sensitivity to light of the outer covering of a primitive creature, and this covering, or a section of it gradually growing more and more sensitive with succeeding generations, can hardly explain the development of an eye.

The eye could only help markedly in adaptation and survival after it had been fully constructed, and after it had been linked by nerve cells to a sensitive optical center in the brain. Any slow progress in the simultaneous development of these various factors would not have been of much assistance in the animal's struggle for adaptation and survival.

Nature employs both inner-plane weather and outer-plane weather to prevent life forms from stagnating. Climatic changes compel creatures to change their structure or their habits, or both, or perish. And the record of the rocks prove that the soul activating life forms has the ability thus to bring about the necessary changes. Using psychokinetic power it moves first in one direction and then in another until it finds the correct answer to its problem, even though the answer be highly complex.

Nor is the compulsion to make progress or suffer confined to creatures lower in the scale than individual man. Due to inner-plane laws, each person has his own inner-plane weather. It is mapped by his progressed aspects. At times this inner-plane weather is favorable, and at other times it is severe. Under progressed squares, semisquares, sesquisquares and oppositions, both the mental outlook and the outer-plane conditions change. Difficulties enter the life that require the utmost effort to surmount.

The changing inner-plane weather insures that the individual will be confronted with one problem after another. If he is to live as he desires, he must find ways of solving these problems. He must develop initiative, intelligence and ingenuity or severely suffer.

The Super Intelligence that permeates all nature, and whose mind has formulated the overall pattern of the universe, has wisely provided that nothing shall stagnate. Even the hardest rocks in time disintegrate. The only changeless thing we observe is change itself.

Within this overall pattern there seems to be opportunity for souls to exercise freedom of choice. They can choose this road or that road, but if the form they occupy strays too far from the broad highway which leads toward the development of those qualities which will be of assistance in realizing God's Great Evolutionary Plan, that form perishes and becomes extinct.

To develop those qualities of character, and those natural aptitudes, which eventually will be of assistance in realizing the cosmic plan, it is necessary that the soul which is being trained to become a useful workman in this vast constructive enterprise should undergo the kind of experiences that will provide that training. Other life forms or men only learn to overcome obstacles, which alone constitutes accomplishment, through encountering obstacles. Each obstacle surmounted adds to the ability to overcome other obstacles in the future. Only through overcoming smaller difficulties does the soul develop the ability to overcome larger difficulties.

The Universal Intelligence has thus provided the necessary obstacles, but not in some whimsical manner, and not as if actuated by human favoritism or prejudice, but chiefly through the inner-plane weather and the outer-plane weather. As we shall see, life forms on earth from time to time had to struggle and make new adaptations to meet the environmental conditions brought through a change of outer-plane weather. And not only is outer-plane weather—as a study and application of Course 15, *Weather Predicting*, demonstrates—influenced markedly by the inner-plane weather, but through progressed aspects the inner-plane weather markedly influences the individual.

The inner-plane weather mapped by progressed aspects brings periods which are lush and easy, alternating with periods in which the individual is called upon to exert himself to the utmost to handle new and adverse conditions. And it is through learning how to overcome the difficulties which thus are periodically presented that the talents of the individual are developed and he makes progress in character.

But in addition to bringing to each soul just the kind of conditions which it most needs to develop the qualities it should gain eventually to fill its particular job as a workman in God's Great Evolutionary Plan, this inner-plane weather has held the progress of life within certain broad boundaries. Life forms that have strayed too far from the broad highway that leads to the development of a form which in itself is a microcosm, or miniature universe, have one after another become extinct. The inner-plane and outer-plane weather have made it possible only for those forms to continue their progress which have led in the general direction of developing the form of man.

Man, as a universe in miniature, is the climax of physical form and function. But he is not the climax of evolution; for by developed self consciousness he has become fitted to continue the evolutionary process indefinitely after physical life amid the far more advantageous surroundings of the inner plane.

The Carboniferous Period

Progress had been such that by the Devonian period of the Paleozoic era there were not only land plants, but flowering plants, and not only land insects but vertebrate land creatures. These laid their eggs in the water and went through the early stages of their lives in the water. The earliest record of such an amphibian is the track of Thinopis antiquitus, found in 1896 by the late Professor Beecher of Yale, and presented by him to the museum.

The two periods following the Devonian frequently are considered together as the Carboniferous period, which commenced with the Mississippian period about 300 million years ago. They are so grouped because both favored vast and luxuriant vegetable growths. In the Mississippian, or Lower Carboniferous period, the climate changed from the more severe Devonian, becoming mild and moist and highly favorable for the growth of rank club mosses in the low-lying swamps. The dominant life at that time were the sharks. They not only were abundant, but grew to immense size. Chiefly they were of the ancient type that fed on shellfish and later became practically extinct. The vegetation of this period in time became coal, and while the first record of an amphibian goes back to the Devonian, the first fossil bones of amphibians were found in the Edinburgh Coal Measures in Scotland which belong to the Mississippian period.

The sixth period of the Paleozoic era, the Pennsylvanian, or Upper Carboniferous period, began about 255 million years ago. Seed bearing shrubs and trees of many species are known to have existed at that time, but their flowers were small and inconspicuous. The mild climate continued, and throughout the swamps the coal plants reached their greatest size. These plants, largely club mosses, were of soft, spongy wood and made rapid growth. They propagated by means of spores, which were carried by the wind, giving a uniform character to the forests all over the world.

These coal swamps formed a suitable environment for spiders, scorpions, centipedes, ancient insects, snails and amphibians. Early in the period insects grew to enormous size. There are fossil cockroaches as long as one's finger, and dragon flies with a 29-inch wing spread. None of these insects, however, had voices, or were much specialized in any direction. The climate and the plants were of a monotonous sameness, and the insects needed little complexity to meet such conditions. At the end of this period, as the climate became more arid, the reptiles developed from the amphibians, as there is direct evidence to show. This was the time in which the great coal deposits were formed through the burial of great masses of swamp vegetation. It is also the period in which the earliest fossil remains of true reptiles have been found.

Once more let us return to the amphibians. As previously mentioned, they must spend the early days of their lives in the water. In this early stage they have gills like fish for breathing in the water, and must secure their food supply there. They are thus well fitted for a climate in which at certain times of the year there are heavy rains that fill the rivers and pools, and at other times of the year there is an arid condition in which the rivers and pools dry up. After their early growth they can live the balance of their lives on the land.

Now toward the end of the Pennsylvanian period climatic conditions in certain sections grew increasingly arid. The pools and rivers not only dried up for part of the year, but it gradually came about in certain regions that they remained dry the year round. Creatures that must lay their eggs and live their early days in the water could find no suitable habitat.

The soul struggling for life and expression was hard put to find some way of meeting and adapting its physical organism to this new situation. Probably innumerable amphibians, trying first this and then that failed. But finally extrasensory perception was able to guide the psychokinetic power in at least one female amphibian to a successful solution of the problem, and she handed on the knack she had acquired to her progeny. The solution took the form of making a more suitable egg.

Additional yolk was supplied to the egg, so that the young might have an ample food supply to nourish it beyond the larval stage and the metamorphosis in which it transforms from a tadpole with gills into an air breathing animal. Henceforth this transformation, which previously had taken place in the water, would take place in the egg before hatching, as it does in the embryo of all higher animals, including man. And to protect the egg better from the inclemencies of the weather, and from drying out, it was provided with a covering, or shell. The gills were lost forever, in so far as the hatched animal is concerned, and no longer did it need to have water in which to lay its eggs. They could be laid in the dry sand and hatched by the sun. Thus the true reptile was developed, many species of which no doubt were adapted to the desert, as they are today, where they live and thrive far from any pond or stream or spring of water.

Cold Weather

During the Paleozoic era fish were dominant. Its seventh and last period is the Permian period which began about 215 million years ago. It was marked by a drastic change in climate. The land which had risen somewhat toward the close of the Pennsylvanian period continued to rise, so that after the arid condition at the commencement of the period, which was drastic in its way, there followed quickly a glacial age even more rigorous in its cold than the one so recently passed, which was successfully weathered by man.

The higher mountains become, the colder the atmosphere. This is the time of the Appalachian Revolution, in which great mountain chains were forced up in the eastern section of the United States. The ice age of early Permian times, however, started in the Southern Hemisphere. But from there it spread over much of the world. These harsh conditions destroyed the old forms of plant life that were ill adapted to the cold, and developed hardier stock. The new plants, starting in the Southern Hemisphere because the cold became severe there first, spread northward. They consisted of cycads, ginkgoes, modern ferns, and conifer trees. These became the dominant plants during the Age of Reptiles which was to follow.

This cold period destroyed many of the invertebrate animals, and they became extinct. Those that did survive have changed little up to the present. It was due to these severe winters that the insects, in order to weather the cold and live through a long period in which there was no food, devised the method by which the larva encases itself in a cocoon, in which it lies dormant throughout the winter, to awaken when the weather becomes warm and emerge transformed from a crawling caterpillar into a mature insect with wings. Insects previous to the Permian period did not undergo this important change. They were compelled to devise some way to endure the long cold seasons of early Permian times, and met the situation in this way.

Just how much intelligence does the extrasensory capacity of an insect have? That is a hard one to answer. But here is an example of the exercise of that intelligence which takes place each year in the mountains near Los Angeles.

The southern slopes of these mountains in early summer are spectacular with the tall candle-like stalks of Yucca plants white with bloom. They belong to the lily family, but depend for pollination on one particular insect, the Yucca Moth. This moth flies by night to a Yucca flower, collects the pollen from its stamens, and kneads this into a little ball which she tucks under her chin. Then she flies to an older Yucca flower and lays her eggs in some of the ovules in the seed box, being careful to leave plenty of ovules free from eggs so that they will mature into seeds and insure future plantings. Finally she applies the pollen from the little ball she carries to the tip of the pistil, thus insuring that the flower shall be fertilized and that the seeds, upon which her race depends, shall grow. Her larvae, when they hatch, live on the Yucca seeds. She thus provides for the continuation of her race; for Yuccas do not bear seeds in regions where there are no Yucca Moths.

Dinosaurs Dominate the Earth

Reptiles first appeared on earth during Pennsylvanian times, developing rapidly and specializing in many different directions during the period of aridity and cold of the Permian. In the layers of rock formed from the mud and sand laid down in the Permian period the presence of not less than five out of the total fifteen orders of reptiles have been discovered; and there is much indirect evidence that six or seven other orders were then present, such as the turtles, beaked lizards, and crocodiles.

Then came a new geologic era, the Mesozoic, which means the age of medieval life. It began about 190 million years ago and is often called the Age of Reptiles, because while both mammals and birds put in an appearance rather early in this era, they were not able to compete with the reptiles but were forced to live inconspicuously. The era is divided into four periods of unequal duration.

The first period of this Mesozoic era was the Triassic period. In the Permian period there were reptiles called cotylosaurs, and from these developed protosaurs. Then in the Triassic period from the protosaurs developed the parasuchia, and finally as modifications of these came the dinosaurs, meaning terrible lizards. They soon grew to be the most formidable and the largest land animals the world has ever known.

Their remains are first found in Middle Triassic formation, representing a time some 175 million years ago. There were other kinds of reptiles during the Mesozoic era, but the dinosaurs dominated all four periods. The imprints of their feet are to be found in the rocks of Colorado, Arizona, Utah and Wyoming as clearly as if they had walked over the muddy ground only yesterday. Some footprints are as much as fifty-two inches long. But all became extinct at the close of Mesozoic times. They were giants, and some were terrible. But as colossal as they were, weighing up to eighty tons, they had minute brains and like that of modern reptiles, their blood followed the temperatures of their environment.

The Age of Reptiles, as the whole of the Mesozoic era is called, was a period of calm lasting perhaps 135 million years. To be sure, land rose and fell in different places, and some of the land went beneath the sea and some of the sea floor became dry land. But it was nothing like the Carboniferous period, during which volcanic action was intense and the atmosphere was humid and stifling, storms were almost constant, rain came down in torrents, and the sun was seldom seen because of the heavy clouds. Nor was it like the drastic cold of the Permian period. The reptiles, including the dinosaurs, were able to meet the problems of the more placid Mesozoic era with amazing success.

The first marked progress of the dinosaurs was probably hastened by a semi arid climate which compelled them to travel long distances for food and water. After they first became established, the climate underwent various changes and in response to these changes some of them developed new forms and new habits. In fact, their habits were quite as varied as those of the mammals at the present time. Some of them were no larger than a house cat, but some, like Brachiosaurus, reached a length in excess of 80 feet. Some fed upon vegetation; but some were carnivorous, preying upon the herb eaters. Practically the complete skeleton of one of the carnivorous kinds, Allosaurus, measuring 34 feet 2 inches in length, by 8 feet 3 inches high, was unearthed from Como Bluff, near Medicine Bow, Wyoming, and is now mounted and on exhibit in the American Museum of Natural History, New York City.

The largest of the carnivorous species so far found is Tyrannosaurus Rex. It is 47 feet long and must have weighed as much as the largest living elephant. It stood 18 to 20 feet high, and the hind limbs, which is as common among them supported the weight of the body, were larger than an elephant's. The great jaws were set with teeth an inch wide and from three to six inches long, and the feet were armed with great sharp curved claws. Being so large it could hardly move with great speed, and probably preyed upon the huge herb eating armored dinosaurs that lived at the same time. These had developed an armor of hard scales, and had horns with which to defend themselves.

There were also smaller carnivorous kinds, such as Ornitholestes, which was unearthed from the famous Bone Cabin quarry in eastern Wyoming. It was only seven feet long, being slender and no heavier than a setter dog. In contrast with the bulky forms, it was fitted for great agility. A common feature of many dinosaurs is that they walked on their hind legs and used their heavy tails as a balancing organ and auxiliary support when at rest. Brontosaurus, one of the herbivorous species, is now mounted in the American Museum of Natural History, measuring 66 feet 8 inches long, with a weight estimated at 38 tons.

The oldest dinosaur relic is found in the rocks of Central Europe; but in a layer of rock almost as old, dinosaur remains have been found in North America. Scientists conclude from this that they probably had their origin in a land mass then existing between America and Europe, where later lay Atlantis. They spread to almost all parts of the world, being found in Southern Asia, Africa, and Australia. What is now Connecticut Valley, late in the Triassic period was an arid region, like that now of the Southwestern United States. The plant life there was rushes, pines and ferns, with no sign of flowering plants. There were dry stream beds that occasionally became flooded with water; and extending down the valley was an estuary. In the mud about this body of water walked numberless dinosaurs, 150 species having been counted.

These footprints, some of which became uncovered, were first observed in 1802, but were thought to be those of birds. Their true nature was discovered in 1835. The mud had changed to brownstone, and a slab containing important records was for 60 years used as a flagstone with its reading matter turned down. This brownstone slab is now at Amherst College, and other slabs have revealed the presence of the 150 different species of dinosaurs mentioned.

For a long time it was a disputed question whether dinosaurs laid eggs or gave birth to their young alive; certain living reptiles following one method and certain other ones the other. The question was definitely settled by the American Expedition to the Gobi Desert in Mongolia where, in 1923, it unearthed 25 fossil eggs belonging to several species of dinosaurs. In several of these, which had been broken, could be discerned the delicate bone of the embryonic dinosaur. Furthermore, it was discovered that these giant reptiles had gizzards for grinding food, linking them more closely to the birds than hitherto had been possible. These eggs were laid during the Cretaceous period, about 50 million years ago.

Returning now to the Triassic period, in the upper layers of rock that were laid down at this time are found the fossil remains of the flying reptile, the Pterodactyl. Up to the present time the intermediate stages have not been discovered as fossils, and like the turtles, the pterodactyls suddenly appear in the strata fully developed, making little improvement later. At the present time there is an amphibian, several species of which live in Borneo, which has developed the ability to volplane to considerable distance. It is a tree frog which has broadly webbed feet, and also webs in the angles of the arms that sustain it somewhat in the air, increasing the distance of its leap. There are also at the present day two kinds of lizards, or reptiles, that glide very well. One, the Draco, is a little lizard whose sides stick out into a pair of wing-like membranes supported by a number of long ribs. This apparatus folds like a parachute, anticipating the invention of man, and being used for much the same purpose.

The flying reptiles of the Upper Triassic period are not thought to be direct ancestors of our birds, although there is considerable structural resemblance between them. But both flying reptiles and birds are thought to have developed from a common ancestor that lived in Permian times. Some of these flying reptiles—and they not merely soared but were capable of sustained flight—were no larger than a sparrow. Other species were the largest creatures that ever flew. One from the rocks of the Cretaceous period, mounted at Yale, measures 18 feet wing expanse, and others are estimated to have had an expanse of 26 feet 9 inches. The best specimens come from the chalk beds of Kansas.

This is not the first instance in the world's history of an animal solving the problem of aviation, which only recently has been solved by man. The insects solved it first in the Paleozoic era, probably actuated by the necessity of escaping from the amphibians that fed upon them. But the wings of the insects, while serving the function of flight, are very dissimilar in principle to those of the flying reptiles, the birds, and the bats. An insect's wing is merely an expanded and greatly modified section of the wall of its body. The wings of birds, bats, and the ancient flying reptiles, as well as those of flying fish, are modified front limbs. That flying fish use this ability to soar to escape enemies—sailing above the water by means of large membranous front fins that vibrate rapidly—may be witnessed almost any day on the trip from Los Angeles to Catalina Island. The first fossil of a flying fish occurs also, as well as the first flying reptile, in the Triassic period.

The Triassic period was one of increasing aridity, which placed a premium on the ability to travel far and swiftly in search of food and water. As a consequence, certain reptiles learned to fly, and a large group of dinosaurs learned to run on two legs, just as today in arid regions modern lizards often tend to this mode of locomotion.

Solving the Problem of Keeping Warm

An event of much greater importance that the development of the huge dinosaurs took place during the glacial period of Permian times, although the first records of it are not found until late in the Triassic period. The cold and aridity not only placed a premium on great speed of movement, but also on the ability to keep warm. Reptiles and fishes are very susceptible to cold. This in large measure is due to their imperfect blood circulation. The heart, unlike that of mammals and birds, has less than four chambers, and the oxygenated blood mixes with the impure blood before being sent to do its work throughout the body. Not only do the reptiles lack a pure blood stream, such as is present in the higher animals, but the heart is incompetent to keep the blood stream active enough to maintain a constant temperature in the body.

When the surrounding environment is warm the reptile or fish becomes warm, and when it becomes cool the fish or reptile becomes correspondingly cool and sluggish. This is a great disadvantage in a period of cold weather, such as Permian times are known to have been. The desire for greater activity, and to keep the body at a constant temperature in the face of cold weather, enabled the psychokinetic power guided by extrasensory perception, and no doubt stimulated by favorable planetary energies, to use hormones and other factors to change the scales of certain reptiles into hair, and to develop that great advantage, a four-chambered heart.

This change was taking place before the Triassic period. In the rocks of the Triassic period in South Africa fossils have been found of a group of reptiles known as Cynodonts, that had discarded the type of teeth characteristic of reptiles, and had developed teeth like those of the modern dog. They well bridge the structural gap between reptiles and mammals. In addition to the incentive for warm blood and warm clothing, brought by the cold of Permian times, in the Triassic period there was increasing aridity, placing a high premium on the ability to travel far for food and water in a short space of time. The more perfect circulation of both birds and mammals enables them not only to endure cold without hibernating, as reptiles do, but also conduces to greater activity.

The first fossil remains of mammals are found in the late Triassic rocks both in Germany and in the eastern United States. These early mammals were small creatures, not larger than a rat. Some had teeth like the present insect eating animals, suitable to living on worms, insects, young reptiles and young birds, while the teeth of others were suggestive of the rat kangaroos of Australia, indicating a herbivorous diet. Although the first mammal records thus date back about 170 million years, the mammals developed little until after the commencement of the Cenozoic era, over 100 million years later.

That the mammals, which seem to have developed from reptiles that traveled on four legs, remained small and inconspicuous creatures for a period almost twice as long as has elapsed since they became important forms of life upon the earth probably was largely due to the inner-plane weather at that time which, as well as outer-plane conditions, favored the dinosaurs. During this period there was an immense number of varied forms of both large and small carnivorous dinosaurs which preyed upon all other living creatures. Only small forms of mammals which could dart into shrubbery, or into burrows in the ground on a moment's notice, could hope to survive the stalking terrors of that day; for no mammal that ever lived, with the exception of modern man armed with explosives, could have been successful in combat with them. It was only when the Age of Reptiles came to an end, and all these monsters mysteriously became extinct, that the mammals had a chance to become important creatures.

The second period of the Mesozoic era, the Jurassic period, commenced about 155 million years ago. It ushered in a mild humid climate in which vegetation grew lush and rank in vast swampy lowlands. As a result of this abundant and easily obtained vegetable food supply, many of the dinosaurs that hitherto had been rapid travelers and many that had been creatures of prey, resorted to the swamps and the easier method of living upon a vegetable diet. They became indolent, living much of the time in the water, and less and less adapted to a strenuous life on land. This possibly contributed somewhat to their extinction at a later period when a severe climate set in once again.

The First Birds

Many of the dinosaurs of the Triassic period ran on two legs, and food seeking and the desire to escape terrestrial enemies caused some of the smaller species to take to the trees. A dinosaur running swiftly over the ground on its hind legs would be helped immensely in its speed by using its arms as does a human runner, if these became broadened to resist the air on the backward swing. This desire for speed may have been somewhat instrumental in changing, through psychokinetic power, the fore limbs into wings.

The species that went into the trees no doubt learned to perch there out of the way of predatory creatures below. They had, judging from their descendants, three long fingers provided with large claws which were used in climbing, even as a modern bird, the Hoactzin of British Guiana, still retains such claws on its wings for this purpose. The pursuit of their prey would lead them to jump rapidly from one limb to another, and if possible from one tree to another.

The desire thus to move rapidly through the trees without descending to the ground, which was full of peril, if intense enough would act upon the astral form of the creature and on the astral body of the offspring, modifying it in the attempt to find a method of accomplishing the sought for results. The scales of the reptile on both the fore limbs and on the rear limbs thus became modified to offer greater resistance to the air. Probably the primitive bird used the hind limbs as well as the front ones for soaring; for in young birds such as squabs there are quills that sprout on the legs so arranged as to indicate that in the remote past their ancestors had steadying feathers on the legs as well as feathers on the wings. In present day birds these quills do not become feathers, but on some early fossil birds just such feathers are present.

This development from reptile into bird seems to have taken place in the arid Triassic period, and as a result, early in the Jurassic period, which we are now considering, there may have been a four-winged bird, although no fossil remains of it have been discovered. The oldest bird fossils so far found were buried in the sedimentation laid down in the latter part of the Jurassic period, something like 135 million years ago. Two specimens of this bird have been found, both from a quarry at Solenhofen, Bavaria. This bird (page 84) is known as the Archaeopteryx.

In many respects it is more of a reptile than a bird. It has teeth in both jaws, for instance. The bones of the hand have not yet fused into a modern wing, and although a wing is present, the fingers are yet armed with claws. The breast bone, which in modern birds must be very strong to support the Right muscles, is as vet quite feeble. In modern birds the tail is short and blunt and the feathers disposed about it fanwise; but this primitive bird has a long lizard-like tail, along either side of which the feathers are arranged in pairs. It appears in the fossil records long after the first record of a flying reptile.

Unlike the primitive mammals, which lived by stealth and so in constant dread of the terrible reptiles that they had no chance to develop, the birds, well fitted to any extreme of temperature by their feathery covering—which in many respects is superior to hair or fur—seem to have undergone a steady progress from the very first.

The third period of the Mesozoic era, the Comanchean, or Lower Cretaceous period, began about 120 million years ago. During this period the larger reptiles that had developed during the Jurassic period became extinct. The world was still replete with reptiles, but they were more specialized and the monsters had disappeared. This result may have been brought about by the change that took place in the vegetable world, although throughout most of the world the climate was still warm and much like that of the previous period. Nevertheless, in Eastern America the ferns, ginkgoes and cynads began to take a secondary place, and true flowering plants commenced to be the dominant form. The Comanchean period, therefore, may be called the period of the rise of flowering plants.

The fourth and last period of the Mesozoic era, the Cretaceous period, commenced about 95 million years ago. The reptiles were smaller than previously, but were numerous and had reached the height of specialization. There were numerous queer forms, some of which were unbelievably grotesque. Toward the end of the period, great mountain chains rose, there was considerable climatic change, and the reptiles were not so diverse in development, many of the queer looking ones having vanished. Here, for the first time, are records of the first placental mammals—those that do not lay eggs—the marsupials, being primitive forms of the opossum and kangaroo.

The records of this period are mostly marine, the chalks of Kansas producing the fossils of birds, as well as those of sharks, fishes, sea turtles, sea reptiles, and fish eating flying lizards. The birds of this period all retained the teeth of their reptilian ancestors. Two mounted skeletons of each of two birds from this period are preserved at Yale. One, Hesperornis, is a diving bird about four and one half feet long that, like our penguin, had lived so exclusively in the water that it had lost the power of flight, although small wings are present. The other is a bird much like our modern gull except that it still had teeth.

Among the dinosaurs there were two divergent races. One race kept all four feet on the ground and developed into the present day crocodiles. The other race moved on two legs and developed into birds. But toward the end of the Cretaceous period extensive land areas were greatly elevated and all the innumerable dinosaurs, for some reason not well understood, became extinct. Whether it was the cold, small mammals that ate all their eggs, or some infectious disease with which they were confronted, they were unable to meet the new condition successfully.

The mention of infectious disease brings up the question of the place of viruses in evolution. They give the appearance of being on the boundary line between organic and inorganic substance. When they appeared first on earth is not known. In fact, as they are invisible with the ordinary light microscope, little was known about them until 1941 when they were first made visible with the electron microscope.

Some live on animals, some on plants, and some are bacteria eaters. Outside the bacterium these latter seem dead, but when they enter the bacterium they quickly come to life, multiply rapidly through sexual union after the manner of most protozoa, and when the colony thus generated numbers several hundred they burst forth through the ruptured walls of the bacterium.

Chapter 6

Development Among Mammals

Serial Lesson Number 130
Original Copyright, 1949
Elbert Benjamine
a.k.a. C. C. Zain

Evolution of Mammals

EVOLUTION OF MAMMALS

EOCENE	OLIGOCENE	MIOCENE	PLIOCENE	PLEISTOCENE
Creodonts Fox-like mammals	Daphaenus--better teeth	Dogs as large as bears	Giant wolves True dogs	Wolves Dogs
Moeritherum 3½ feet high long nose	Palaeomastodon larger, 4 tusks	Dinotherium Tetrabelodons 4 tusks smaller than elephants	Tetralophodon 4 tusks, jaw 6 feet long Mastodon, Imperial Elephant	Mastodon Mammoth Colombian Elephant. Imp. Eleph.
Eohippus 4 toes front, 3 behind Orohippus 13 inches high Epihippus middle toe dominant	Mesohippus 3 toes in front, 3 behind, size of coyote Miohippus larger	Merychippus 3 toes front, 3 behind, size of pony Protohippus teeth developing like modern horses	First modern horse, still with "dew claws"	Horses Asses Zebra
Protylopus size of jack rabbit	Size of sheep several branches	Protomeryx hoofs like deer. No teeth missing Procamelus teeth reducing	Pliauchenia more advanced	Giant Camels
Pelycodus Lemurs. Notharotctus	Propliopithecus	Dryopitheous Sivapitheous	Propliopitheous Australopitheous Paranthropus	Man Apes Monkeys

EVOLUTION OF MAMMALS

EOCENE	OLIGOCENE	MIOCENE	PLIOCENE	PLEISTOCINE
Small fleet mammals	3 groups small mammals horns develop	Larger in size more complex in structure	Abundant Rhinoceroses	Rhinoceroses
Mammals size of rabbit	Small hoofed mammals	Larger and more complex struct- ure	Several Bisons	Giant Bisons Ox
Mammals size of rabbit	Small hoofed mammals	Deer-like, an- telope-like, but primitive	Sheep, antelope Deer	Deer Sheep Prong-horn
Mammals size of jack-rabbit	Primitive pec- caries, primit- ive giant pigs	Giant pigs as large as a horse Peccaries	Peccaries	Peccaries Pigs
Small mammals size of coyote, without pronounced proboscis	Proboscis de- veloping	Larger size, longer pro- boscis	Primitive Tapirs	Tapir
		Porcupine-like animals First bears	Many porcupines Cave-bears of large size	Porcupine Bears Cave-bears
Creodonts Fox-like mammal	Claws developing	True cats Sabre-tooth tiger	True cats Saber-tooth tiger	Lion, Tiger Cats

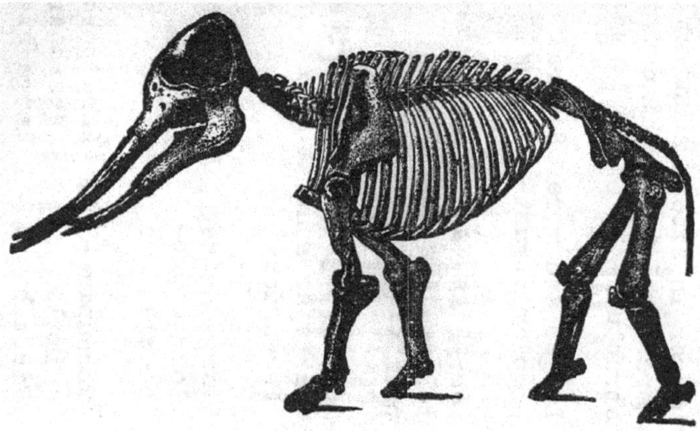

Skeleton of Mastodon

Chapter 6

Development Among Mammals

ONLY a portion of the mammal species that existed a few million years ago have survived to this day. Yet the species that have survived are so numerous that any adequate history of their development would require several volumes. The various stages of progress of some kinds are as yet unknown; but there are a vast number which have left their bones at successive stages of development where they were covered by mud, sand, dust, or volcanic debris, to become fossils.

These constitute an indelible record by which modern mammals may be traced back unerringly through their genealogical chain to more primitive stock; link after link being revealed as new remains are uncovered, until the chain disappears in the primitive creatures of the Triassic period some 170 million years ago. Here, however, limitations of space will permit me to mention, and most briefly, the progress of only a few forms. And for that purpose I have selected a dozen species of mammals with which everyone is somewhat familiar, for they can be seen at any time on the farm or in the zoo: the dog, lion, elephant, horse, camel, ape, rhinoceros, porcupine, bear, pig, deer and bison.

It should be understood that only a few small and insignificant primitive mammals existed up to the commencement of the Cenozoic era about 55 million years ago. These minute creatures were the forerunners of modern marsupials. Their remains are found in the Triassic of North Carolina. These Allothera, as they are called, are also found in the Triassic of Europe and the Jurassic of South Africa.

That they must have had a terrible time trying to survive is indicated by the record that for 115 million years, more than twice as long as since their enemies, the dinosaurs, disappeared and gave them a chance really to progress, they remained little, sly creatures which were constantly on the dodge to keep from becoming food for the big lizards that dominated the world. But as soon as the dinosaurs disappeared they started rapid progress. And it may be assumed that the conditions under which they lived, with their lives in danger every moment, previous to this time had sharpened their wits and given them considerable intelligence. The dinosaurs, on the other hand, are noted for the minuteness of their brain in comparison to their brawn.

Even before the extinction of the dinosaurs, the little ancestors of modern mammals had made some progress, for in the Cretaceous period of about 90 million years ago there are found the fossils of numerous minute marsupials which had developed from the older stock. And apparently contemporaneous with them, as indicated by their fossils, were also the first placental mammals, the undoubted ancestors of all modern forms. But all of both types, until the dinosaurs vanished, were constantly forced to hide, and given no opportunity to develop size.

The mammals of today all are of the placental type, with the exception of the opossums of both Americas, the Coenolestes of Ecuador and Colombia, and certain primitive types which live in Australia. In the Triassic of South Africa a group of reptiles has been found which very closely approaches the form of the most primitive mammals. Three such primitive mammals, which have made little progress above the mentioned advanced reptiles, survive in Australia even to the present time. They are the duck mole, the spiny anteater, and the proechidna, which is an anteater without a vernacular name.

These mammals have pouches in which, like the marsupials, they place their young. But as do the reptiles, and unlike the marsupials, they lay eggs. They have never reached the stage of caring for their young by retaining them in their body until they are mature enough to make the hazards of external life less dangerous. The eggs they lay are large and contain much yolk. In the case of the duck mole, after the eggs hatch there are no teats for the young to suck, but the milk oozes out of numerous pores on the front of the mother's body and is licked off by the young. In the case of the anteater the egg is placed in a temporarily developed pouch and when it hatches milk oozes out into the pouch.

These mammals belong to the group lower than the marsupials. They are known as Monotremes. They seem to represent the transition stage between lizards and placental mammals. It appears that the desire to protect their young from the rigors of difficult environment had stimulated extrasensory perception and psychokinesis, and that psychokinesis guided by extrasensory perception had actually taken a long step to attain this desire; for, unlike the young of reptiles, their young are provided nourishment, the forerunner of true milk, until they are husky enough to forage for themselves. But as yet they have not provided the fuller protection that the placental mammals have devised for their offspring.

Neither have the marsupials, although they have made a decided advance in this respect over the monotremes. They no longer lay eggs, but give birth to their young. But unlike higher orders of mammals, the young are born prematurely, so helpless in fact that they cannot even suck. Thus, the marsupials have only partially solved the problem. The mother takes the young one in her mouth and places it in the skin pouch and adjusts its mouth over the teat and then injects the milk down its gullet. A special arrangement of the glottis in the young enables it to breathe while milk is thus being pumped down it, without danger of strangling.

The kangaroo is the most familiar marsupial of today. With the exception of the dingo—a dog presumably reaching there through a human agency—all the mammals of Australia are monotremes or marsupials. This is due to the fact that Australia was cut off from the balance of the world before the placental mammals developed. In Australia there are marsupials that have solved most of the other problems solved by various species of more developed mammals. Some are similar to squirrels, some to our wolves, some to our hoofed animals, some to our burrowers, etc. Being isolated from the rest of the world, the development of mammals in Australia and the adjoining islands has been along independent lines. The other continents having repeatedly been connected by land since placental mammals came into existence, have forms of life with much more in common.

That marsupial mammals and placental mammals solved the problem of adaptation to similar environmental conditions along almost identical lines, even though the problem could have been solved in different ways, was probably as much due to extrasensory perception and the inner-plane weather favoring this particular solution, as to the outer-plane conditions.

The Placental Mammals

The placental mammals, to which group man belongs, made a distinct advance over the marsupials. Premature young are under a great handicap. The death rate due to enemies and climatic conditions is high. It is a decided advantage that the young shall have reached an advanced stage of growth at birth, and thus soon be able to endure severe weather and follow the mother about, or perhaps remain hidden by her without perishing.

This desire for greater protection for the young brought psychokinesis into play in the development of the placenta; a tissue that links the unborn young and the mother in a prolonged partnership. By means of this tissue the blood vessels of the young are brought into close contact with the blood vessels of the mother, and thus absorb from her dissolved nutritive matter, oxygen, water, and the necessary salts; also giving to her in return the dissolved waste products. One of the most important functions in this long sleep of the unborn is that it enables it to be born with a well-developed brain.

Even to glimpse the development of the various present day species of mammals from more ancient and primitive types it becomes necessary to know something of the interrelations of land areas throughout the world at the times when important mammalian modifications were made. I will, therefore, make brief mention of the changes in land areas that took place at different times. Due to shrinkage of the earth, and to the impact of inner-plane weather, land areas rise and sink. There are periods in which portions of continents sink below the sea, islands and other land areas disappear, and land bridges that previously have connected continents subside leaving no way open for long lapses of time by which land creatures or plants from one continent can find their way to another.

There are various ways of accurately determining the existing land areas at any period of the past. For instance, the shell fish and other marine forms that are able to migrate only along a coast line and never across deep bodies of water, are entirely different along the warm Asiatic coast near Japan from those occupying the coast along the Arctic Ocean. When land rises across Bering Strait, making a complete land bridge between Asia and America, the Arctic waters no longer flow southward along the west American coast. Instead, the waters are warmed by the Japanese Current, and the shore life from the Japanese region spreads along the continuous shore line following the land bridge, and is found all along it and well down into California.

When this land bridge subsides, and instead of leaving a small shallow strait, as today, the Arctic Ocean is given free access to the North Pacific, these cold waters exterminate the Asiatic form of coast life that are suited to warm waters, and we find the fossils of the purely Arctic types. There is then no longer a route of migration open between Asia and North America, and the shell fish of the Arctic follow the cold waters well south along the western coast of North America. Thus, without taking the space to go into detail, it will be understood that when I speak of land connections between continents being made or broken, or of land areas being raised or lowered, I am not referring to tradition, but making the assertion upon sound and fully accepted scientific opinion.

In the Cretaceous period, something more than 55 million years ago, it is quite certain that all the continents of the world were connected by land; for dinosaurs, which were strictly land animals, have left their fossils in the non marine Cretaceous rocks of every continent. Because there were both marsupial and placental mammals during this period, it is possible that the latter also found their way to all the continents, although no fossils of placental mammals belonging to this period have thus far been found in Australia or South America. South America was then connected with Australia by land across Antarctica, which then had a mild climate; and there is strong evidence that this connection, and a land connection between Brazil and West Africa, existed into Basal Eocene times.

Eocene Period

The first period of the Cenozoic era (meaning era of modern life) was the Eocene, which commenced about 55 million years ago. It is usually divided into two sections, the lower being called Basal Eocene, and the upper merely Eocene. In Basal Eocene times both Europe and Asia were well connected by land at the north with North America. South America was then cut off from North America, and continued thus cut off well into the Cenozoic era, the Pacific and Atlantic being connected across what is now Central America. The Culebra Cut of Panama, through formation belonging to the Eocene period, reveals marine shells which would not have been present if it had then been a land area.

The climate during the previous period, the Cretaceous, was very warm even into the Arctic regions; warm climate plants being found in Greenland and Alaska. In Eocene times it was somewhat cooler, but still genial, as shown by the innumerable remains of great crocodiles and large palm trees in Wyoming and Idaho. North America in Basal Eocene extended further to the east than today, and Florida and the North Gulf Coast were submerged; but otherwise this continent was in shape very much as now. The Appalachian Mountains, older than the western ranges, were already worn down almost to a plain with a few peaks sticking up, as peaks thus stick up at present in large areas of Texas, the highest being in North Carolina. The Rocky Mountains and the Sierra were in existence, thought much lower than today, and the eastern coast and the interior were vast plains. The continent was largely forested; willows, poplars, sycamores, oaks, and other modern trees being mingled with conifers. England was temperate in climate.

Lower Eocene rocks of northwestern New Mexico, and Upper Eocene rocks of eastern Wyoming, South Dakota, and Montana, have yielded numerous fossil mammals; but for the most part they are types that later became extinct, leaving no descendants. All are small, no mammal of say 50 million years ago as large as a sheep ever having been found. The ancestors of all modern mammals, then very small and primitive, had their origin probably in Cretaceous times on land well toward the North Pole, according to Wortman as a result of his studies of the fossils at Yale. From thence, as the then sub tropical climate cooled, they were forced south into Europe and North America, where they displaced more primitive forms that were less adapted to environment than are those of today.

In Eocene times proper, there was a submergence of the Atlantic Coast and Gulf Coast of North America, and the Gulf extended up the Mississippi Valley as a long arm of the sea. On the Pacific another long, narrow arm of the sea extended up the great valley of California to Oregon and Washington. The Mediterranean covered most of Southern Europe, the Pyrenees, Alps, and such great ranges not yet having been lifted. Europe was completely separated from Asia by a strait east of the Ural Mountains joining the Mediterranean with the Arctic Ocean. America and Europe were well joined by land, and the wave of mammal migration coming down into both continents from the north gave them mammals that were more nearly identical than at any time since. This is particularly true of the lowest formation of the Eocene proper, where innumerable fossil mammals have been found. Before the close of the Eocene, as shown by the Uinta formation in Green River Valley, northeastern Utah and northwestern Colorado, the mammals had become totally different from those of Europe; for the land connecting these two continents had sunk below the sea.

In Basal Eocene the carnivorous mammals were represented by the creodonts; nothing similar to the dog family or the cat family yet having evolved. These creodonts were flesh-eating mammals, with teeth very much like cats, but with heavy tails and blunt claws. The first known member of the dog family in North America is a very small fox-like creature living about 40 million years ago found in Upper Eocene. It is believed to be a direct descendant from the primitive creodonts. It also appeared in Europe at about the same date. Furthermore, about this period there is found in the Libyan Desert of Egypt an animal about three and one half feet high. It is the Moeritherium which, although having no proboscis as yet, shows by the nasal openings the beginning of the development which finally supplied the elephant with a trunk. Fortunately, the intermediate links between this short faced creature and the elephant of today, to which it is the ancestor, have been found; showing a gradual development in size and length of proboscis.

The horse, on the other hand, had its origin, and most of its development in America. Eohippus, the dawn horse, is found in the very earliest Eocenes laid down about 55 million years ago, in Wyoming and New Mexico. It was a small graceful creature twelve inches high at the withers, with an arched back. It had much the proportions of a fox terrier, except that the feet were already beginning to be modified from toes into hoofs. The hand had four complete toes, each with a hoof-like nail. The foot had three such toes, although a splint shows where its ancestor had another toe which had atrophied. Animals which require speed run on their toes, not flat footed. This lifts the outer toes on either side from the ground somewhat, tending to place the weight on the middle and longer toe. Thus the middle toes become stronger and better adapted to carrying the weight, and fortified against the impact of the ground; while the outer toes, not getting much use, fail to grow.

In Middle Eocene, about 10 million years later, is found in Wyoming and New Mexico, Orohippus, the mountain horse. By that time this ancestral horse had developed in size to thirteen and one half inches tall. The splint remnant of the fourth toe had been entirely lost, the outer finger of the hand had been shortened, and the teeth had been modified. In Upper Eocene, another five million years later, is found Epihippus, still larger than Orohippus, yet with three toes behind and four in front, but with the middle toe of each front foot becoming quite dominant. From the vestigial structures of these early horses it is quite certain that they descended, as did all present day mammals, from a five-toed ancestor. The five toed horse, however, while frequently mentioned, so far has not been found.

Strange as it may seem to some, North America is the original home of the camel, where he remained and developed for some 35 million years, only migrating to the Old World in Pliocene times. Before the Upper Eocene the camel is not to be distinguished from other small primitive mammals. But in Upper Eocene distinguishing characteristics were far enough advanced that in Protylopus, a creature the size of a jack rabbit in the Uinta formation, we can recognize the ancestor of modern camels, the intermediate steps being represented by a very complete series of fossils.

The Primate mammals, to which apes, monkeys, and man belong, there is much evidence to show, developed from the same small insect eating mammals of the Cretaceous period from which the carnivorous mammals developed. The descendants of this primitive Cretaceous insect eater diverged into two branches: those which were strictly flesh eating became the carnivorous mammals, and those which adopted a strict diet of nuts and fruit which took them into the trees, developed into the Primates. The remains of these first Primates are found in the Lower Eocene of both Europe and America, and the transitional forms between the insect eaters and the Lemurs are found in North American Basal Eocene of about 50 million years ago. The first recognized Primate, such as Pelycodus, is found in the Wasatch formation of the Great Basin, some 45 million years old. In Upper Eocene formation of the Green River Valley, the Lemur, Northarctus, is abundant. These lemurs, so abundant in the northern hemisphere during the Eocene, have since that time been found only in Madagascar, tropical Africa, and southern Asia, where very similar lemurs still exist at present.

The earliest tapir is a mammal the size of a coyote, found in the Wasatch formation of the Eocene of the Great Basin of North America, of about 45 million years ago. Most of the development of the tapir was accomplished in North America. It migrated into Europe during the Oligocene period, about 25 million years ago, but only reached South America where alone it exists as a much larger animal today, in Pleistocene times, less than a million years ago.

Found in the Lower Eocene of North America, of about 50 million years ago, is a small animal that in the course of several million years developed into a huge creature with two horns on its nose, resembling in many respects the rhinoceros. This creature, the Titanotheres, very numerous at one time, is not the ancestor of the rhinoceros; for it became extinct. Why it died out no one knows; perhaps from disease, as there was no predatory animal at that time large enough to menace it. It is believed, however, that the true rhinoceroses had their origin and much of their development in North America. The earliest form is a small fleet mammal without the characteristic horn developed, from the Wind River formation of the Eocene period, some 45 million years old. From this developed a number of distinct forms, increasing in size as time rolled on, and finally becoming extinct in America in lower Pliocene, about three and one-half million years ago.

Those animals which have an even number of toes like the ox, deer and pig, are called artiodactyls. The oldest members of this group so far found are from the Lower Eocene of the Great Basin. One of these little creatures the size of a rabbit is the ancestor of such mammals as chew the cud, as do sheep, bison and cattle. Another, somewhat larger, is the ancestor of the pig. Most of the development of the artiodactyls took place in the old world, to which they migrated very early.

Oligocene Period

The second period of the Cenozoic era is the Oligocene, commencing about 35 million years ago. It was marked by the rising of a land bridge between America and Europe, across which mammals migrated both ways. This is shown by the close similarity of the mammals found in the White River deposits of the early half of this period in northeast Colorado, western Nebraska, eastern Wyoming and South Dakota, with those of Europe. The American climate was warmer than at present, although the gradual cooling that finally in the Pleistocene brought a glacial age set in during the second half of the period. In Europe the climate was sub tropical, being even warmer than Eocene times; palms growing in Germany. The latter half of the Oligocene witnessed a great change; for the land bridge between Europe and America again subsided.

As a consequence the mammals of the two continents, as shown by the John Day deposits of eastern Oregon, being isolated from each other, and not able through interbreeding to disseminate to each other their developed characters, developed each along lines that were demanded by peculiarities of climate. Hence, while early Oligocene deposits reveal mammals in America and Europe that are almost identical, by late Oligocene the mammals of the two continents had so developed along divergent lines as to be quite dissimilar. Fortunately, Oligocene formation, both early and late, yield vast quantities of fossils.

In North American Oligocene formation there have been found many different kinds of primitive cats, some of which undoubtedly were the ancestors of modern lions, cats, leopards, etc.; but none of which had developed into a near likeness of these. The claws, however, in some species had become well developed, and two broad groups were emerging, one with teeth gradually becoming more and more like those of present day species, and the other with the upper canine teeth greatly elongated. These became the sabre-toothed group, the most terrible cats of all time, and the dominant beasts of prey down to less than a million years ago.

In Oligocene times the dog group had quite distinctly separated from the cat group, although both certainly had a common ancestor in the creodonts of the previous period. The dog, represented by several species of Daphaenus, the largest dog of that time, was hardly larger than a coyote. The backbone greatly resembled that of the cats, and the claws were somewhat retractile, which is a cat feature. The teeth were small and had only partly developed the shearing edges characteristic of modern species, but the skull was dog-like, rather than like that of the cats.

In both Africa and India the descendants of the more primitive elephant of the preceding period had developed in size, in the height of the skull, and in the length of the proboscis. As found in Oligocene deposits its lower jaw had elongated and short tusks had developed both in the upper jaw and in the lower jaw. It, therefore, had four tusks, instead of two as in modern species. This Oligocene ancestor of the modern elephant is called Palaeomastodon.

In North America, in the lower Oligocene formation is found Mesohippus, a horse which now had developed to the size of a coyote, and in somewhat later formation to the size of a sheep. It no longer had four toes in front, as did its Eocene ancestor, but had three toes in front and three behind, the middle toe having developed to much greater strength and the missing toe being represented by a splint. In Upper Oligocene, representing several million years later, is found Miohippus, which is very much like its ancestor, Mesohippus, except that it is much larger.

The camel of the lower Oligocene, about 35 million years ago, as shown by the White River deposits of North America, had developed to the size of a sheep, although more slender and fleet of foot. In the John Day deposits of eastern Oregon, representing Upper Oligocene some 15 million years later, the camels had developed into several branches: the giraffe camels, the gazelle camels, and the two other groups which ultimately became the llamas and the true camels.

The Primates—monkeys, apes, etc.—became extinct in North America at the close of the Eocene period, although they existed in South America, and still continue there. They also became extinct in Europe during the Oligocene period, continuing their development in Southern Asia and Africa. The transitional form between Northarctus of the Eocene and the manlike apes was discovered in the Oligocene of Egypt. It is Propliopithecus, which lived about 30 million years ago, and greatly resembles the gibbons of the present day.

There are a great many fossils of primitive rhinoceroses in the White River beds of North American Oligocene. There were three different groups, one of which was light and fleet of foot, another which was massive and slow, and still another which was the ancestor of modern species. Some of these had horns, and others had none, but they were larger than their Eocene ancestors, although not nearly so large as rhinoceroses at present. Tapirs also occur, larger and with longer proboscises than previously.

In this John Day formation, belonging to the Oligocene period, are also found primitive peccaries, and primitive giant pigs. There are found in this formation, likewise, innumerable small hoofed mammals that were developing into deer-like, ox-like, and sheep-like forms. The ox, nevertheless, mostly developed in Europe.

Miocene Period

The third period of the Cenozoic era is the Miocene period, commencing about 19 million years ago. The Atlantic Coast and Gulf region were again submerged, a cool current from the north driving out tropical forms along the eastern coast of America and replacing them with northern forms. On the west coast the sea again extended up the California Valley, leaving the tops of the Coast Range as small islands, Miocene formations in California being as much as from five thousand to seven thousand feet thick. There was a good connection between America and Europe, but the bridge across Bering Sea was broken. Also, for the first time in the Cenozoic era, early in Miocene the principal elevation of the Coast Range took place, and the Sierra and Plateaus of Arizona and Utah were elevated higher than previously. It was warmer than at present, although not warm enough for crocodiles to be found far north. It was even warmer in Europe, due to the continent being broken and intersected by warm arms of the sea, the climate being like that of India. It was also a great period of mountain making in Europe.

This period is chiefly notable for the marked development both in size and number of the mammals. Many kinds of mammals in the Miocene period grew to much greater size than their present day representatives, and there were far more genera and species than now. Practically all families of mammals now on earth are represented by more primitive Miocene forms, and many groups then prevalent have since become extinct. It was the period of greatest mammalian abundance, there being great numbers of hares, mice, pocket gophers, squirrels, marmots and beavers, as well as numerous larger animals.

There were many kinds of dogs, some of which were as large as any existing bear; truly formidable beasts. There were also true cats, developed from the more primitive forms of the Oligocene period, but they were not so large as the lion and tiger of today. The sabre tooth tigers were likewise present, having made notable advances, but not yet so large as the huge beasts that became extinct at the close of the ice age. The fossil remains of weasels, martins, otters, and raccoons have been found; but these too are represented by more primitive animals than those bearing the same name today.

The elephants of the preceding period had continued to increase in size, and had developed along several lines. One, the Dinotherium, found in Europe and India, had lost its upper tusks, and instead had a lower jaw that with its tusks bent abruptly downward. The four tusked mastodons increased in size and in tusk development, one genera, the Trilophodon found in France, not nearly so large as the Indian Elephant, had an enormously lengthened lower jaw. It represents an intermediate stage toward the development of the modern elephant. This was the first elephant to reach America, coming by way of Asia in early Miocene times, about 18 million years ago. The members of the four tusked mastodon group are called Tetrabelodons.

The considerable elevation of land areas and the growing aridity of North America during Miocene times caused much of the forested regions to lose their trees and shrubbery and become broad plains covered with coarse grasses. Animals, such as the horses, that previously had fed upon the succulent browse of the trees, were compelled to turn to the harsher and harder food found on the plains. Also, due to the drying up of the rivers, it became necessary to travel immense distances for water. The result of this was a pronounced modification in the teeth and in the feet. There were a number of horses in Miocene times. They had increased in size to that of small ponies. Most of them were three toed, although the outer toe scarcely touched the ground, the middle toe bearing almost the whole weight, being much larger and a well formed hoof.

Merrychippus, a horse of the Middle Miocene, about 15 million years ago, is the intermediate link between the browsing type and the grazing type of horse. The milk teeth in this species are short crowned and have little cement, as in previous horses; but the permanent teeth are higher crowned and quite heavily cemented. In Upper Miocene, developed from Merrychippus, is found Protohippus, still having three toes, but with teeth much more similar to those of the modern horse.

The camel of Lower Miocene times, Protomoryx, had made a decided advance over the camel of the previous period. For one thing, it is larger. All the teeth are present, which is not the case with modern camels; and the two toes, instead of having cushions as at present, are armed with sharp hoofs like those of a deer. The teeth also indicate a change to make them suitable for grazing rather than for browsing. The first tooth reduction is shown in Procamelus, a descendant belonging to the Upper Miocene, showing changes in the shape of the foot and various other desert adaptations.

Descending from the apelike creatures of the preceding period, Sivapithecus, found in India in Middle Miocene deposits some 15 million years old, while still an ape, has numerous manlike characteristics. From this common stock, about 18 million years ago, the Orangutan branched off, and since that time has developed away from its human characteristics. The Chimpanzee and the

Gorilla branched from the common stock in Middle Miocene, some 15 million years ago. Instead of developing human characteristics, they developed other qualities, the Chimpanzee becoming less human than previously, and the Gorilla retaining its structure about such manlike characteristics as were common to the original stock in Middle Miocene times. Since Middle Miocene times the living apes have mostly been developing along lines almost the opposite of those followed by the ancestors of man.

The bears are descended from the same ancestors as the dogs, the intermediate links having been found. Nothing that had developed far enough to be called a bear is present before late Miocene times. Its fossils are found in Europe, and it did not reach America until the next period.

The history of the porcupines is also quite well known. They, however, underwent their development in South America, where today no less than six families and 29 genera are known. Some are no larger than a rat, some are large; in fact the largest living rodent, the Wart Hog, belongs to this group. Some live in trees, some in water, and some in the ground, and some have long prehensile tails that they use as an extra hand. It was not until Miocene times, when a land bridge was formed between North and South America, that porcupines are found in North America, and then only a single species, as now, represented by the present day Short-Tailed Porcupine.

The rhinoceroses of Miocene times were abundant both in North America and in Europe, and show considerable advance in size and structure over those of the preceding period. The ox and most of the deer underwent their development chiefly in the Old World, but the prong horn antelope and the Virginia deer have their ancestors well represented in American Lower Miocene, and can be traced accurately from that time to the present. Pigs were numerous in Miocene times, but for some reason not clearly known, but which no doubt was influenced by inner-plane weather, the giant pigs, huge creatures as large as a horse, did not persist beyond the Miocene but, together with many other mammals, became extinct.

Pliocene Period

The fourth period of the Cenozoic era, the Pliocene period, commenced about seven million years ago. It was cooler than the previous period, and a complete land connection existed between North America and Asia across what is now Bering Sea, giving a marine fauna on the Pacific Coast like that of Japan.

Early in the Pliocene the true wolves developed, some of immense size, and by Middle Pliocene, about four million years ago, the modern genus of dogs had become established. Many cats are found in the formation, some of which are very large; both such cats as became lions and tigers, and those of the sabre tooth group. Of the elephant group of this period, now practically as large as modern genera, there was one, Tetralophodon, found in Nebraska, with four tusks, and a lower jaw six feet long. Another type, found over most of America, is Dibelodon, very much like a mastodon except that its teeth were not so developed. The mastodons, differing from the elephants in their teeth and in some other features, also were present in Pliocene times; and in America the Imperial Elephant that then roamed the hills about Los Angeles and had a very wide range, was larger than any present day species.

The land bridge between America and Asia enabled horses and camels to migrate from America to Asia, and enabled elephants, now becoming somewhat similar to modern species, to migrate to America. The first of the modern horses is found both in America and Asia in Upper Pliocene, about two million years old. The feet have one toe each, but the splints are also prominent enough to be "dew claws." The camels of this time are somewhat more advanced in structure than those of the previous period, being represented by a mammal called Pliauchenia.

Bears reached America during this period, and the native pigs, called Platygonus, were numerous and more highly developed in many respects than the modern American descendant, the peccary. There was a great abundance of rhinoceroses, four or five kinds native to America, and several others that migrated in Miocene times from the Old World. Several different kinds of ox and sheep are also present and the prong horn antelope and Virginia deer; but the moose, caribou and wapiti developed in Europe and did not reach America until the next period.

Nothing has been found of the Primates in North America belonging to this Pliocene period, with the exception of a single tooth found in Nebraska of a manlike ape, or an apelike man, called Hesperopithecus. In Bechuanaland, 80 miles north of Kimberley, in South Africa, late in 1924, Dr. Raymond Dart, professor of anatomy at the University of Witwatersrand discovered in a limestone cliff which at the time was thought to date well back in the Pliocene, the remains of a manlike ape, which he called Australopithecus Africanus (Southern manlike ape of Africa). Some, even yet, are inclined to place this find back as far as the beginning of the Pliocene, seven million years ago, basing their opinion chiefly on geological evidence. But in the 24 years since he made this find Dr. Dart, and Dr. Broom, chief paleontologist of the Transvaal Museum were able to find fossils representing possibly 15 individuals. And chiefly because of their high intelligence, as they were using fire, Dr. Dart now wishes he had named these Transvaal pygmies, which are believed by some to be the ancestors of modern man, Homunculus (little man). And according to an article by him in the Autumn, 1948, issue of American Journal of Anthropology, he places them contemporaneous with the Men of Java, who lived at the commencement of the Pleistocene period, about one million years ago.

Pleistocene Period

The fifth period of the Cenozoic era is the Pleistocene period which commenced about one million years ago. Due to considerable elevation in land areas at the end of the Pliocene period, the climate of the entire Northern Hemisphere was greatly cooled, ushering in the glacial age. The Pleistocene is the age of ice during which the ice sheet came down from the north reaching as far south as 40° latitude, and over New York attaining a depth of 10,000 feet. During the 970,000 years of the Pleistocene the ice came down no less than four distinct times, each time again receding. Between these intrusions of ice there were long interglacial periods in which the climate usually grew even warmer than it is now.

At the commencement of the Pleistocene period had there been a primitive explorer roaming over America he would have found all the southern part of what is now the United States covered with such spruce and pine as at present grow in Canada. Had he gone north through this thick forest, he would have been stopped abruptly by such a wall of ice as is now to be seen when approaching Antarctica. It extended from what is now New York City to the State of Oregon, covering all of Canada and at least half of the United States.

In addition to the elevation of land areas, climatic changes are also influenced by the relation of the earth to the sun. Not only by the variation of the inclination of the earth's axis to the ecliptic, but also by the distance of the earth from the sun. The earth's orbit fluctuates considerably. It was elongated some 100,000 years ago, then was almost circular about 50,000 years ago, became elongated again in 20,000 B.C. and now is once more moving toward the circular. It will reach its nearest circular shape in about 20,000 years. Except as influenced by terrestrial conditions, the weather gets colder when the orbit elongates, and warmer when it is more circular. Therefore, we are now moving toward a warmer climate which will be at its hottest about 20,000 years from now, when it is expected tropical vegetation will grow on the shores of the Arctic Ocean. But, according to George Gamow, professor of physics at George Washington University, large masses of ice will again start creeping down from the north, by about the year 50,000 A.D., as the fifth glacial advance, and will completely cover all the cities of Canada and northern United States.

The animals of the Pleistocene period, though different in size and in other respects from those of the present, would readily be recognized as the general group to which they belong if seen today. The American Mastodon ranged the forests, and of the true elephants there were three species: the Mammoth, the Colombian Elephant, and the Imperial Elephant. There were giant wolves, and there were true cats, some of large size, as well as the Sabre Tooth Tiger, which was more massive than any living tiger.

There were camels that were considerably larger than those of the present. There were also several species of true horses, with a single hoof on each foot, and there were bison that were much larger than any existing species. Moose, caribou and wapiti had reached America from Europe, and there were huge cave bears in various parts of the world. Furthermore, there were huge ground sloths, and armored glyptodonts, and a number of other mammals that have since become extinct. But in addition to these extinct forms there were those with which we are familiar, such as the peccary, mink, weasel, martin, skunk, otter, badger, wolverine, raccoon, fox, coyote, puma, etc., all present in America. And as there were rigorous climatic changes, the competition between forms and the struggle to survive must have been exceedingly intense.

Such a brief survey of the development of mammals is not altogether satisfactory, yet at least I trust it shows conclusively that all existing forms have developed from more primitive preexisting mammals. In many cases complete series of fossils have been found showing the chief steps by which changes in structure were made, so that to doubt their genealogy is to doubt the evidence of one's eyes.

A study of these fossils indicates that whenever members of a single species separated over a long period of time with an impassable barrier between such as the destruction of a land bridge between two continents, or in some cases by a desert, or a broad river, or a mountain range; that the members of each region developed along different lines. No longer able to commingle and breed, and thus impart developing qualities each to the other; being in a different environment, they each develop qualities called for by that environment. Both in structure and in habit they become further and further apart, until they become distinct species.

We have witnessed the creation of breeds by artificial selection; for all the tame pigeons—fantails, pouters, carriers, etc.—are known to be derived from the rock pigeon; and all domestic chickens are known to be descendants of the Jungle Fowl of India. There are instances, also, within the knowledge of man of new species being the result of isolation. Rabbits, for instance, were turned loose by mariners on certain islands off the southwest coast of Europe several hundred years ago. These rabbits, although of the same stock as rabbits on the mainland, are today a totally different species, quite incapable of breeding with the rabbits of the mainland.

Chapter 7

Development of Man

Serial Lesson Number 131
Original Copyright, 1949
Elbert Benjamine
a.k.a. C. C. Zain

Transition to Upright Posture

From left to right are the skeletons of the gibbon, orang, chimpanzee, gorilla, and man, showing transition to the upright posture.

The gorilla, man's nearest living kin.

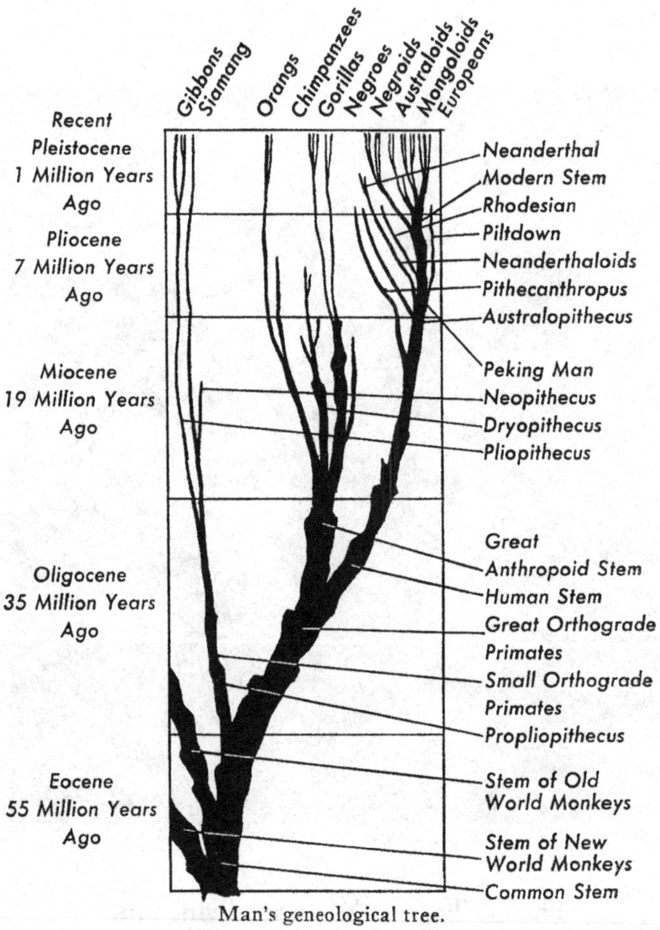

Man's genealogical tree.

Chapter 7

Development of Man

IN Chapter 3 (Serial Lesson 127) the chief characteristics of the vertebrate animals were mentioned. Man possesses all of these characteristics. Furthermore, as his young are born alive and nourished by mammary glands, he is a mammal. Of the mammals, the method by which the young are nourished before birth places him among the placentals. These are classified into four groups: the clawed ungiculates, the hoofed ungulates, the finned cetaceans, and the nailed arborial primates. Obviously he has neither the structure and claws of a dog, the hoofs of a horse, nor the fins of a whale. But he does have the flattened nails, the clavicle, the simple stomach, the thoric mammary glands, and all other characteristics of the primate group.

The ancestors of all Primates, as stated in Chapter 6 (Serial Lesson 130), there is good evidence to show, were small insect eating mammals that lived in the Cretaceous period some 75 million years ago. The first fossil Primates are found in the Washatch formation of the Great Basin of North America, some 45 million years old. Such is Pelycodus; but the transition between the earlier insect eaters and the true Primates is found in Lower Eocene formation, about 50 million years old. They had adopted a strict diet of fruits and nuts which took them into the trees. Still later, in Upper Eocene, the lemur called Northarctus, which resembles present day lemurs, is abundantly found in the formation of the Green River Valley in America.

The lemur is often referred to as a "half ape"; for the brain is not developed so extensively as in the true monkeys, and the second digit of the foot still bears a claw, instead of the nails which the other digits carry. It is thus much more closely related to other mammals than are the apes and monkeys.

The present day Primates are divided into several families. The marmosets belong to the family Hapalidae. The Capuchians, howler monkeys, spider monkeys, etc., belong to the family Cebidae. The Old World Monkeys, baboons, macaques, etc., belong to the family Cercopithecidae. The anthropoid, or manlike apes belong to the family Simiidae. Man is classed in yet another family, the Hominidae.

The nearest relatives to man, structurally and biologically, are the anthropoid, or manlike apes. There are four living genera of the anthropoids, and several extinct genera. The living genera are the Hylobates, or gibbons; the Simia, or orang; the Pan, or chimpanzee; and the Gorilla, or gorilla. The gibbons and orang at the present live in the Orient, and the chimpanzee and gorilla are natives of Africa. The transitional form between Northarctus of the Eocene and present day anthropoids, Propliopithecus, was found in the Oligocene formation of Egypt dating back about 30 million years.

The orang branched off from one anthropoid genealogical stem in Miocene times, about 18 million years ago. Somewhat later, in the Middle Miocene formation of India, representing an age of about 15 million years, there has been found the fossil of an extinct form, Sivapithecus. It lived rather close in time to another form, Dryopithecus, and not far from the time when both the chimpanzee and the gorilla branched off as different trends in anthropoid evolution.

To sum the matter up, we may say that the gibbon of today is somewhat less manlike in its characteristics than its ancestor, Propliopithecus, which lived about 30 million years ago. The orang of today is less human in structure, and probably in its habits, than its ancestor of some 18 million years ago. The chimpanzee is less manlike today than its ancestor of 15 million years ago, and the gorilla of today has at least no more likeness to man than its ancestor of 15 million years ago. We have much reason to believe, therefore, that there has been a constant widening of the structural breach between man and the living apes for not less than 30 million years.

The Most Recent Discovery of Primitive Man

The Java fossil man (Pithecanthropus) and the China fossil man (Sinanthropus) go back a long way, but the fossils of the oldest man discovered to this date (1949) are those of certain pygmies of South Africa. In late 1924 such a fossil was dug from a limestone quarry about 80 miles north of Kimberly. The discovery was announced by Dr. Raymond Dart, professor of anatomy at the University of Witwatersrand. This man ape, as it then was called, was only a baby, not more than four years old, that had died in a limestone cave which slowly filled up with stalagmites, turning the skull into rock. Happily the entire skull was preserved, and Dr. Dart at the time stated: "It is a creature well advanced beyond modern anthropoids in just those characters, facial and cerebral, which are to be anticipated in an extinct link between man and his simian ancestors."

He named it Australopithecus Africanus, meaning southern manlike ape of Africa. And some geologists placed its age as going back to the commencement of the Pliocene period, some seven million years ago. But later digging has convinced Dr. Dart that the pygmies, of which this find was a member, often entered limestone caves in their hunt for baboons. And in the 25 years since this first find, Dr. Dart has kept persistently on the trail of this possible ancestor of modern man, and has unearthed numerous fossils of it, some quite mature.

As a consequence of these later finds he says in the autumn, 1948, number of the American Journal of Physical Anthropology that now he wishes he had named it Homunclus, meaning little man.

These pygmies lived earlier than either the Java man or the China man. They lived on the treeless savannas of what is now central Transvaal. They had apelike faces, were about four feet high, and weighed around a hundred pounds. But in spite of their small size, their brains were almost as big as that of the oldest man previously discovered, the Java man, who stood around five feet eight inches.

Says Dr. Dart in the 1948 article: "These intelligent, energetic, erect, and delicately proportioned little people were as competent as any other primitive human group in cavern life made comfortable by the use of fire, in the employment of long bones as lethal weapons, in the cunning and courage of the chase and in internecine strife."

In 1947 Dr. Dart's diggers unearthed the back part of the skull, and a lower jaw from an immature pigmy. Near this skull were the skulls of many baboons which had been bashed in from above or behind with a club. The inference is that these pygmies hunted baboons by blocking all but one exit of a baboon cave colony, then as the apes ran out clubbing them from the side of this exit. As these pygmies lived at least 100,000 years, and perhaps many hundred thousand years before the China man, previously the earliest known man to have used fire, and as charred bones give conclusive evidence that they used fire, he named this new man Australopithecus prometheus.

After Dr. Dart's first find, others also took up the search. And in 1936, Dr. Broom, chief paleontologist of the Transvaal Museum in Pretoria, found several fragments of an adult which he has now renamed Plesianthropus transvaalensis. Then two years later he located, from some rocks casually picked up by a schoolboy, a third genius of the same subfamily as Australopithecus, which he named Paranthropus robustus, meaning manlike and strong. During 1948 Dr. Broom was able to locate the fossils of at least 12 and possibly 15 individuals of Plesianthropus, representing the adult from head to toe. The plants and animals with which the finds of both Dr. Dart and Dr. Broom were associated were definitely those of the early Pliocene. If they were contemporaneous that would mean that the South African Man Apes lived about seven million years ago. But Dr. Dart now doubts this, and tentatively places them at the end of the Pliocene or the commencement of the Pleistocene, only about a million years ago.

No doubt time and research will reveal not only the direct line through which modern man descended from Propliopithecus, but the significant successive steps. At the present moment the finger points at Australopithecus as the most promising ancestor, from one of the three genera of which, perhaps from Australopithecus prometheus, present man has evolved. The evidence is conclusive enough that present day man descended not from a monkey, but from an animal ancestor which was also the ancestor of the apes. But as to our immediate anthropoid ancestor, and just where on earth man developed, the evidence is as yet not conclusive, and must await the unearthing of new fossils.

The theory is that Australopithecus developed the erect posture and increased brain activity due to the section of the continent in which he lived becoming arid. But no doubt inner-plane weather conditions contributed to his progress as much as his outer-plane environment. Approaching aridity, as we have indicated in preceding chapters, is more often than not the cause either of extinction or the development of new structural adaptations by which extinction is escaped.

In a forested country, life is easy, food abundant, and there is little urge for swift development. This may be one of the reasons the gorilla, living in dense forests far to the north of Australopithecus, has remained in development where he was 15 million years ago. But in a region growing arid, food diminishes, water may be had only in certain places, predatory beasts have a great advantage, and the struggle for life becomes strenuous. In such a hard existence, wandering from place to place in search of food, yet beset by numerous enemies, the development of intelligence would prove the best adaptation. The erect position would free the hands for the examination of objects. Such examination is the foundation of all knowledge.

There is a lot of mystical rubbish in print that places man on earth back in time hundreds of millions of years. And it may be that evidence in time will prove that some species of Australopithecus lived well back in the Pliocene period. But there is no evidence that man lived on earth prior to the Pliocene. There are eoliths, which are flakes of flint that some believe were roughly chipped by human agency and used as primitive weapons and implements, that certain scientists have thought might be a bit earlier than the Pliocene. But other scientists not only believe these eoliths assumed their present form through natural agencies, such as being struck by boulders rolling along a stream bottom, but they doubt that they belong to a time earlier than the Pliocene.

According to Dr. Dart's finds, however, man had made quite a development by the commencement of the Pleistocene. But before discussing the development of man since that time, during the last million years, let us consider some of the factors which show the close kinship between other animals and man. The proof of such kinship, which would require a volume for full discussion, rests upon at least the six following different and separate lines of research, any of which alone would be considered sufficient in a court of law to establish it as a fact.

Animals Have Extrasensory Perception

Before taking up indications of man's kinship to animals, as I have repeatedly made reference to different life forms possessing in some degree both extrasensory perception and psychokinetic power, I believe I should elaborate somewhat on this repeated statement.

All my life I have spent as much time as I could spare from work in close contact with wild life, and have had opportunity to observe on occasions what seemed to me to be good examples of their extrasensory perception. But in reference to the kinship between animals and man it seems advisable to give conclusive instances in which animals have employed extrasensory perception.

Quoting from the November 15, 1945, Church of Light Quarterly Report:

"Even those customarily alert are at times caught napping. In the thirty years he has taught two Brotherhood of Light classes a week in Los Angeles, until this summer, Elbert Benjamine has missed working only one day on account of ill health. That was in the spring of 1922 when for a day he suffered from the 'flu.' But recently, during the second half of 1945, he has had a period of illness. It occurred merely because he failed to take the indicated precautionary actions for his progressed aspects. In his zeal for Church of Light affairs he permitted his adrenaline and cortin supply to become exhausted. In addition, war restrictions prevented him from obtaining his customary diet. But now he has recovered and is back on the job."

The illness started July 22, 1945, under progressed Mars sesquisquare Uranus *r* and *p*, and progressed Moon square Mars *p*, reinforced by minor progressed Mercury opposition Mars *p*, and released by transit Jupiter conjunction Uranus *r* (birth chart in Chapter 3, Serial Lesson 41, Course 1, *Laws of Occultism*).

On August 22 I was taken to a hospital. From the time I was carried on a stretcher from my home, my dog Duke (birth chart and progressed aspects in Chapter 3, Serial Lesson 127) crouched in corners, whined almost continuously, and refused to eat. On the third day, between 10:00 and 11:00 a.m., all at once his behavior changed. He trotted about the room, wagged his tail, and in his dog way begged for something to eat. This sudden change in attitude was so marked that my wife, Maria M. Benjamine, at once left him and went to the mail order department, close at hand, where she reported how he was acting. Rev. Edward Doane at once said, "Duke knows Elbert is going to get well." Maria then returned to our living quarters. Within ten minutes after she arrived there the telephone rang. The head nurse at the hospital, a mile away, told her over the phone that the doctor had asked her to call Mrs. Benjamine and inform her the crisis was past, and that Mr. Benjamine would certainly recover.

In *The American Weekly* for December 5, 1948, Dr. J. B. Rhine, Director of Parapsychology Laboratory, Duke University, relates a similar event, but in which the event was more drastic.

He tells of a cocker spaniel which watched the illness of his mistress in an old Georgia mansion. For days, exhibiting grief continuously by his demeanor, he stayed in the deserted bedroom beside the bed from which his mistress had been taken to a hospital.

He did not whimper, but mutely retained a sad expression. Then one afternoon, with nothing unusual happening in his vicinity, all at once, and quite unexpectedly, he started the mournful howl which dogs frequently give in the presence of human death. The servants in the house, giving it this common interpretation, at once said their mistress had died. Checking later with the hospital it was found that the woman did die at approximately the time the cocker started howling.

In the same article Dr. Rhine tells of a man in Canada who left his German shepherd dog penned up securely before starting on a train trip which he anticipated would keep him from home some time. His orders were that the dog was not to be released until he returned.

Then something unexpected came up and the dog's master returned much sooner than he, or those he had left in charge of his estate, had expected. But he had to alight from the train several miles from his home at a flagstop. In the meantime the dog had chewed his way to freedom, and much to the owner's surprise—for no one but himself knew he would get off at this flagstop at that time—he was met there joyously by his dog.

This is an instance of either prevision, or of the dog getting the information telepathically from his master's mind. But, apparently, dogs also at times pick up the experiences of other dogs at a distance. In the same article Dr. Rhine tells of two dogs belonging to the same owner. One of the dogs became ill, and was taken to a veterinarian hospital. The other dog, a Boston terrier, had been in good health. Yet a few days later he was struck suddenly with convulsions as if suffering severely with abdominal pains. Very quickly he returned to his normal health again. But it was learned that his dog friend had died of violent convulsions at about the time the Boston terrier had unaccountably suffered temporarily a similar seizure.

There are a great many instances on record in which both dogs and cats exhibit great fear when in a room where a ghost is supposed to make visits. In the mentioned article such an experience is quoted from Dr. Walter F. Prince's book, *Human Experiences*. In this instance a Newfoundland dog of pronounced fighting propensities, in the presence of such a ghost, deserted the little girl it was his duty to guard and fled whining, with his tail between his legs, to the cellar where he hid in a coal bin; something he had never done before.

The scientists at Duke University investigate dogs that are reputed to have telepathic or extrasensory perception. And they have found some with quite extraordinary telepathic powers. One dog, Pikki, performing with a Russian circus, they gave a series of such telepathic tests. Pikki would do almost anything his master commanded him to do mentally. Screens and other devices were used to make it impossible for the dog's master, Professor Bechterav, to give the dog any kind of sensory clue as to what he wanted him to do.

Thus is there evidence obtained by trained scientific experimenters that dogs on occasion possess the ability to gain information telepathically.

As there is ample evidence that man's personality survives the tomb, the question next arises, does his kin, the animals, survive physical dissolution?

Many have recorded that a pet dog or a pet cat after it died had some time later been seen as a ghost in the locality to which it had become accustomed before its death; moving through the house, jumping up on an easy chair, or doing something else that it habitually did while alive in the physical.

Mr. Pierre Van Paassen in his book, Days of Our Years, relates his experience of a ghostly dog. He was in France sitting by the fire. It was about eleven o'clock at night. Suddenly he felt cold. He went down stairs to throw some coal on the fire, and as he returned something brushed his leg. He looked back and saw it was a large black dog. He had never seen it before. He was surprised; for there seemed no way for a dog to get into the house. It scampered down the steps. He turned all the lights on and made a thorough search of the house. But no dog. So he went to the front door and unlocked it, all the doors and windows being shut and bolted, and called in his two police dogs.

The next night, and on several subsequent nights, he saw the same black dog again, but could find no way by which it could get into, or out of, the house. Then he was sent to Rumania on an assignment for five weeks. When he returned he found his servant, who was already sleeping elsewhere, leaving. Her reason was that she would not work in a haunted house. She said at night a big black dog pushed her door open and came into her room.

Mr. Van Paassen got a man and his nineteen-year-old son to watch with him. They were armed with a club and a pistol. At eleven o'clock they all heard the dog's footsteps as he came running down the stairs from the upper story. All three ran into the hall, and there at the foot of the stairs was the black dog, which calmly stared back at them. One of the men whistled to the dog, which wagged his tail in friendly fashion. But when they started to descend the stairs toward the dog, its figure began to grow hazy and dim, and long before they had reached it, it had completely vanished.

On an evening subsequent to this, instead of the two men, Mr. Van Paassen got his two police dogs to help him watch. When the patter of the black dog's feet were heard overhead the two police dogs pricked up their ears. Then the hackles on their necks became erect and both backed toward the exit door, growling and showing their teeth. Their master on this occasion, however, could not see the black dog. But apparently the two police dogs did; for they let out howls as if in pain, and began snapping and biting as if they were in a terrific fight with something. Mr. Van Paassen could not see what they were fighting, although he stood ready to take part with a stout club. But suddenly one of his police dogs let out a dreadful wail and dropped dead on the floor. The other one backed into a corner whining, whimpering and quivering. The astral dog had won the fight.

But dogs are not the only animals whose ghosts have been reported by reliable witnesses. Major General R. Barter, of the British army, a young subaltern in the Indian service, had an experience in 1854 with a phantom rider on a pony. In 1888 the General sent the story to the Society for Psychical Research, and it is recorded in the Proceedings in Vol. V, Page 459.

Barter, when he arrived for duty in the Punjab, rented a house which had been built a year or two before by another officer who had died some six months previously. One evening some friends had visited him and after he had gone up the trail some distance with them when they left, he turned around to go back to his house, accompanied by his two dogs. As he turned, he heard the ring of a horse's iron shoe on the rocks. Watching in the direction of the sound, presently he saw in the bright moonlight a man riding a horse, a very strange figure; for attended by two grooms, he was in full evening dress, with white waistcoat and high silk hat. The horse was a strong hill pony, dark brown in color, with a black mane and tail. On either side of the pony's head walked one of the grooms. One of the grooms had his back turned and the other's face was hidden by the pony's head. But with one hand each held the bridle close by the bit and rested the free hand on the rider's thigh as if to steady him in the saddle.

The path they were on led only to Barter's house, so he shouted to them, but received no reply. When they came close to him he shouted at them angrily, for he was exasperated that they paid no attention to him.

To this there was no response, but the group halted, and stood motionless. Then Barter recognized the rider as Lieutenant B_______ who had built the house he was living in; for he had previously known him in the service. When Barter sprang to lay hold of the Lieutenant who had been dead six months, the whole group vanished. And he noticed that his dogs, who never strayed far from his heels, had taken to the underbrush.

The next morning he went to visit Lieutenant Deane, who had been in the same regiment with the man now dead. Among the questions he asked the Lieutenant was where the man now dead got his pony. The Lieutenant seemed startled, and said, "Why, how do you know anything about this? You haven't seen B_______ for two or three years and the pony you never saw. He bought him at Peshawar and killed him one day riding in his reckless fashion down the hill to Trete."

Proofs of Man's Animal Ancestry

Within man's anatomy are nearly 200 vestigial structures. These are structures, such as the familiar vermiform appendix, that have served a useful purpose in lower forms of life, but which now serve man no useful purpose. He is in the process of getting rid of them. They have diminished in size, and although yet a part of his physical inheritance, he is discarding them as fast as his system can make adaptation. Thus, they are small and inconspicuous. It seems strange if man is a special creation that he should be encumbered with nearly 200 structures that not only serve no useful purpose to him, although they do serve other animals a very useful purpose, but actually, in many cases, such as the appendix, which I cite because all are familiar with the fact that it is easily infected and causes so many surgical operations, encumber his activities and endanger his life.

In man's skeleton, muscles, nerves, nerve centers, alimentary canal and its tributary glands, the glands of internal secretion, and the respiratory and circulatory systems, he parallels in structure and function other mammals. The few differences, such as the erect attitude and the placement of the skull on the spinal column to permit this attitude (picture on page 122), and the enlargement of the brain, are specializations developed due to certain habits that have been adopted by man and not by other animals, even as other genera of animals differ from each other because of their differing habits.

It also seems strange if man is a special creation that in the development of the human embryo it is first like an invertebrate, then like a fish, then as development proceeds partly reptile-like and partly bird-like, then like a mammal, and finally like a man. The heart when it first develops has a single chamber like the heart of a fish. The auricle next divides in two, giving three chambers like the heart of an amphibian. Finally the ventricle divides and we have the four-chambered heart of the warm-blooded animals including man.

The blood as it develops also passes through the stages of its ancestral evolution. The first red blood cells are large and nucleated like the blood of fishes and amphibians. They then take on the characteristics of the blood of reptiles, and finally, before birth they lose this reptilian structure and become non nucleated and bi concave, the characteristic of the human blood.

Definite proof also is forthcoming as to man's nearest kin among the animals. Blood transfusion affords one such test. The blood serum, or fluid in which the blood corpuscles are carried, of animals of close blood kin when transfused mixes without injury; but the blood serums of animals not closely related is poisonous one to another. The blood serum of the horse is not injurious to the blood corpuscles of the donkey; the blood serum of the hare is not injurious to the blood corpuscles of the rabbit; and the blood serum of the wolf is not injurious to the blood corpuscles of the dog. But the serum of a horse is poisonous to the blood corpuscles of a dog, and the blood serum of a wolf is poisonous to the blood corpuscles of a rabbit. The immunity of the blood corpuscles of one animal to the serum from another undoubtedly is due to such animals having diverged recently from a common parent stock and therefore the serum and the corpuscles have undergone almost no modifications since the divergence occurred. Where the divergences from a common ancestor is so far in the past that the serum and the corpuscles of the different stocks have undergone considerable modifications they become injurious to each other.

The blood serum of man is poisonous to, and destroys the blood corpuscles of other animals, but does not injure the corpuscles of the anthropoid apes. Certain contagious diseases common to man, likewise, are not possible to other animals than the anthropoid apes.

Still, a further test of man's kinship with the anthropoids has been discovered in the "precipitin" test of blood relationship. It has been discovered that if a fresh blood serum of any animal is injected into the veins of a rabbit there will be produced in the rabbit's blood an antibody. Now if into blood taken from the same species of animal originally used to develop the antibody in the rabbit's blood a few drops of the drawn off blood of the treated rabbit be introduced, a white precipitate is formed. If the blood used for the experiment is not of the same species, but of a closely related species, there will be a small amount of the precipitate formed; the amount being determined by the closeness of the relationship between the animal used to procure the antibody and the animal used to get the precipitate.

Thus a scientist, without being made aware of the identity of the animal from which the blood was drawn to form the antibody, and without knowing from what animal the blood was drawn for the test, can determine to what extent they are blood kin. If a horse is used to get the antibody, another horse's blood will show a strong precipitate, indicating a very close relationship; while a donkey's blood will yield a precipitate, but less in amount, showing a relationship not quite so close.

When human blood is used to get the antibody, and other human blood is used for the precipitin test, the precipitate is pronounced. When anthropoid ape blood is used for the test there is a precipitate, but less marked than when human blood is used. And when monkey blood is used for the test there is still a precipitate, but not so marked as when anthropoid ape blood is used. When the blood of other animals is used for the test there is no precipitate. Thus, is the relationship between man and the anthropoid apes established by the almost identical chemical composition of their blood streams.

Physical man and the anthropoids, however, as we have seen, branched from the same parent stock not less than 30 million years ago, and man has evidently been specializing in brains almost ever since. The freeing of his hands due to his erect posture enabled him to hold things before his eyes for examination. This examination reacted upon the brain and made him more anxious to examine other objects; hands, eyes and brain each helping the other to develop. As the hands became more dexterous, the use of implements was discovered, and the thumb then rapidly developed.

In the same strata with the Ape Man of Java (Pithecanthropus), who lived something like a million years ago, possibly contemporaneous with, but by some scientists thought later than, Australopithecus, have been found flints that may have served him as crude implements. This point, however, is open to controversy. Australopithecus, however, who lived at least at as early a date, according to late finds, used both weapons and fire. The weapon which is found with his bones is a club which had a ridged head, the distal end of the humerus bone, with which he bashed in the skulls of the baboons he killed for food.

The China Man, or Peking Man, was unearthed in China near Peking. He lived perhaps 900,000 years ago. His chief claim to fame is that he is supposed to have had a larger brain than the Java Man, and had reached the fire using stage of culture.

The Piltdown Man, Eanthropus, found in England, belongs to still another genus of man. He lived perhaps in the First Interglacial Interval, something over 800,000 years ago; but the precise time is uncertain. The brain case, brow, and back of head are distinctly human of a very low type, the brain capacity being about that of the lowest living savage. The lower jaw and dentition are ape like, the chin being lacking. He walked erect, and in the pit where he was discovered have been found a rough flint spearhead, a stone hide dresser, and a hammer stone. It is thought that he possessed fire.

Probably belonging to the Second Interglacial Interval, some 600,000 years ago, is the Heidelberg Man. His stock branched off from the main human stem at a much earlier date than that of the Piltdown Man. His jaw is remarkable in size and strength, being quite ape like, but the teeth are distinctly human. There was no chin, and no processes for fastening the muscles usually used in speech, therefore, it is doubted if he had language. Crude stone implements found in the same stratum indicate that he was a giant in size and strength; for modern man would find these tools too heavy to work with.

The first man of which we have the entire skeleton—although the fossil pieces of different individuals that have been found represent the complete skeleton of Australopithecus very well—is the Neanderthal Man. He probably descended from the Heidelberg Man, living as far back as the Third Interglacial Interval, at least 250,000 years ago, and up to the end of the last Glacial Period and slightly later, say up to within 25,000 years ago. He is the proverbial Cave Man; for he sought shelter from the cold, and protection from the Cave Bear and other extinct beasts, by taking refuge in caves, where in Europe numerous remains of him have been found. He had great beetling eyebrow ridges, massive jaws, was short of stature, and had a shambling gait. Nevertheless, he had a large brain, was a skilled worker in flints, buried his dead with an outfit for their long journey, and was a believer in magic. In 1921 an African species of this Neanderthal Man was found in Rhodesia, Africa. He is called the Rhodesian Man.

In fitting together the jigsaw puzzle of the time in which each of the mentioned primitive men lived I have presented it in apple pie order. But it is probable future finds will upset both the timing and relationship factors markedly. But as stated, it gives a fairly consistent picture of where, so far as at present known, each of these pre human men fit into the puzzle. Neither the Piltdown Man of England, the Heidelberg Man of Germany, the Neanderthal Man of Spain and France, nor the Rhodesian Man of Africa left any descendants. They were genera of the pre human race that became extinct. All the present day races of the world belong to a single species of a single genera—as shown by their anatomy—that branched into three distinct races, or subspecies, soon after the Piltdown Man branched from the same main human genealogical tree.

Present Day Subspecies of Man

Each of these three subspecies has certain well-marked characteristics. One of the most easily determined is the hair, which not only differs in superficial appearance, but has a different structure when examined under a microscope. All the races of the world at present are thought to belong to one of the following groups:

1. The Polynesian-European Group: These have Wavy hair, fair skin, and long heads. They practically encircle the earth, occupying most of Europe and Northern Africa, and extending in a band through Southwestern Asia and Western America north of the equator along the shores of these two continents.

2. The Asian-American Group: These have Straight hair, yellow or red skin, broad heads and narrow eyes. They occupy Eastern Asia and Western America north of the equator along the shores of these two continents.

3. The Australian-African Group: These have Wooly hair, black skin, and decidedly long heads. They occupy Australia and Africa south of the equator.

As I mentioned, the people of these three groups began to diverge from each other some 700,000 years ago. First the Australian-African group branched from the main stem. Then, perhaps 600,000 years ago, the Asian-American group separated from the Polynesian-European group. Since this branching into three subspecies, the Polynesian-European group, which embraces all of the White race, developed fastest along special lines. The Asian-American group, embracing the Mongols and American Indians, developed not quite so fast, and not in just the same way. The Australian-African group, embracing the Negroes and Australian Black Fellows, developed still less rapidly than the other two branches. On the outskirts of the vast areas of the earth's surface inhabited by each of these three great groups are to be found other people, differing more or less from these marked types. It is supposed that these people, lying on the fringe of well-inhabited areas, have been subjected to certain periods of considerable isolation from the main body of humanity, and have undergone considerable modification due to local environment. They are merely outlying groups of the three main groups which have become highly specialized in some direction.

I have already spoken of the jigsaw puzzle of the relation of the pre human people of the earth to each other and to modern man. But its pieces are nowhere nearly so difficult to fit together as are those of the pattern showing just where present day man had his origin and from what precise pre human people he descended.

He did not descend from the Neanderthal or any other known ancient man. Neither is there any evidence where he attained the development and culture that he possessed when he first appeared upon the scene of the world's records. The first record of the modern type of man is the Cro-Magnon. Just as soon as the glaciers of Europe receded far enough that the climate and conditions were in a measure endurable, the Cro-Magnon came on the scene. No one knows where he came from, but he was a big fellow, many of the skeletons showing him to be 6 feet, 4 inches in height. In intelligence and physique he was the equal of any race living today. His skeleton was like that of modern man, except that he was unusually wide across the face at the cheek bones. He had much culture, being possessed of rough implements of stone, and he drew pictures of animals on bone and on the walls of his caves. The Basques of the Pyrenees of France and Spain are supposed to be present day direct descendants of these Cro-Magnons.

He is also sometimes called the "Reindeer Man" because the reindeer which he had hunted over the steppes of Europe was one of his two chief food supplies, the other being the horse. He killed off the Neanderthal Man, who had no chance against his great intelligence, and possessed himself of their caves. In certain remote and secret caverns he built clay models of animals, and in connection with his religion went through magical ceremonies over these effigies; this magic being, it is thought, for the purpose of overcoming these creatures when he met them in the flesh.

He made very fair paintings of them too, on the walls of his caves, using a number of natural pigments, such as red and yellow ocher. The animals usually so drawn are the bear, the reindeer, the aurochs and the bison. Where did he develop? In lost Atlantis? He first appeared, so far as there are records to show, about 30,000 years ago.

Then about 12,000 years ago there came a new influx of people into Europe. And no one knows where they came from. But the climate by that time had become suitable for agriculture, and the people who then arrived were of the Polynesian-European group. They had long heads, fair skin, and wavy hair.

Previous to this there had been an invasion of negroid people, for their skeletons have been found in a European cave along with those of the Cro-Magnon. But this new invasion, probably the first wave of the dark white race usually called the Mediterranean Race, came to stay, bringing with them polished stone implements, and high culture in many respects, including the use of domestic animals and the cultivation of such grains as wheat, barley and millet. They had pottery, used milk, and lived a life differing very little from that lived a hundred years ago by the majority of European peasants.

But from whence they came, or from whence came later the wave of light Whites, called the Nordic Race, or where either developed their culture, there are as yet no records to show. At some period they set up huge rough monuments of stone; and similar monuments of about the same are found in India and America. Did this custom have its origin in an astrological cult developed in Atlantis?

There was an influx later of still a third type of white man. He is stocky in build, low crowned, and wide between the ears. His is the Alpine Race. He contrasts strongly in disposition and in his political ideals with either the tall, rangy, large boned, high crowned, lone headed, blond Nordic Race, or the slim, smaller, long headed, moderately high crowned brunet Mediterranean Race.

The Alpine has no governing capacity, has great vitality, little organizing ability, great endurance and plodding patience. The Mediterranean is intellectual, philosophical, musical and patient, but lacks the initiative of the blond Nordic. The Nordic is a politician, loves adventure and action, is a natural ruler, and is capable of high cooperation with his fellowman. These three types are to be found more or less segregated in Europe today, and constitute a great political problem; as the three types find it difficult, due to different religious, political and temperamental bias, to unite in a common effort for the good of all.

It should not be thought that any one of the something like fifteen great races into which ethnologists divide the three great subspecies of mankind are pure strains. There is an interblending, a shading of the White group into the Yellow and Black, and of the Yellow into the Black. Mankind, being of a single species interbred freely. Different species when they breed produce sterile offspring, as is the mule, which is the offspring of mating a male donkey with a female horse. But the interbreeding of any of the human races produces fertile offspring.

This interbreeding of races has gone on almost constantly in the past. There is no such thing as a pure White Race, or a pure Yellow Race. Professor Dixon of Harvard University collected and measured all the human skulls he could procure, and collected the measurements made by other anthropologists. Many White skulls showed traces of both Black and Yellow, and many Yellow skulls showed traces of Black and White admixtures, and many Black skulls showed traits derived from Yellow and White ancestors.

Which of these races is superior? Time alone will show. Each has certain natural aptitudes not possessed in equal degree by others. Certainly the color of a man's skin, or the race to which he belongs, is no criterion of inferiority or superiority. Which is superior remains to be demonstrated; for whatever the color of a man's skin, or the character of his hair, we can judge his superiority only by one standard. That standard of superiority is the degree in which he can and does contribute to the welfare of mankind.

Today, with greater facilities of travel, the human species instead of diverging, is amalgamating more and more. More and more the Black Races are being mixed with the White and Yellow, and the Whites are being mixed with the Yellow and Black. Of the White Races the Alpines intermarry with the Mediterraneans and Nordics and the Nordics intermarry with the Mediterraneans and the Alpines.

Nowhere is this fusion being more rapidly accomplished than in the melting pot called America. But where any of these people had their origin or developed their culture previous to about 12,000 years ago there is at present no irrefutable evidence to show.

Chapter 8

Development of Knowledge

Serial Lesson Number 132
Original Copyright, 1949
Elbert Benjamine
a.k.a. C. C. Zain

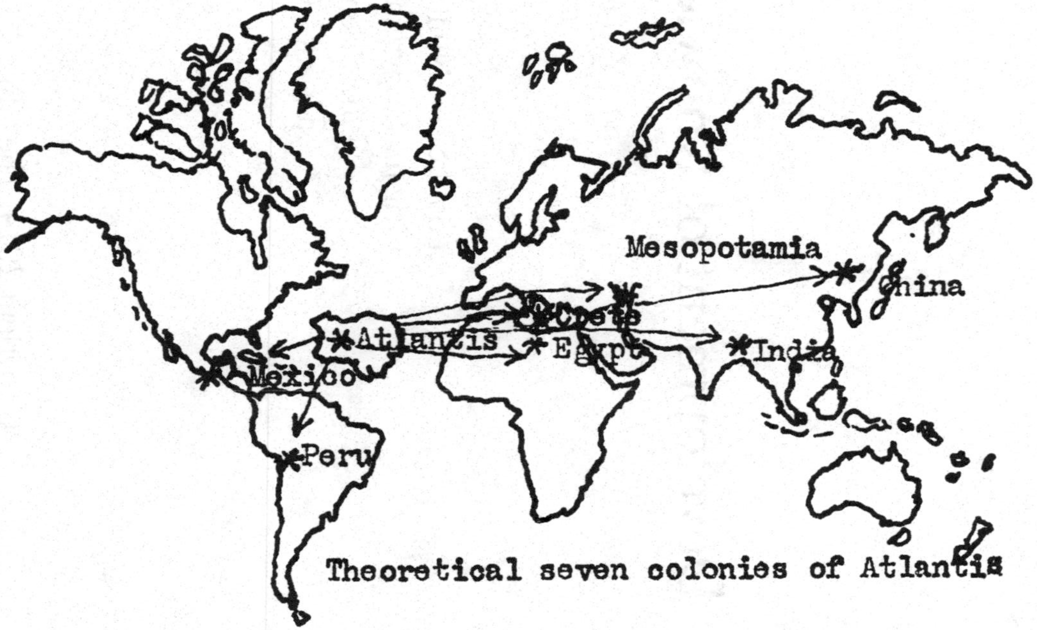

Theoretical Seven Colonies of Atlantis

Sphinx and Pyramid of Gizeh, Egypt

Indian Pictographs

Indian pictographs near Coso Hot Springs, Inyo County, California. Photographed in January 1933, by Fred H. Skinner, who from Nov. 2, 1932, until he passed to the inner plane in 1940, was Vice President of The Church of Light.

Chapter 8

Development of Knowledge

THE first record of the modern type of man is the Cro-Magnon. He appeared in Europe about 30,000 years ago. From whence he came, or where he developed his culture no one knows, nor from whence came either the Dark Whites or the Light Whites that much later followed him with a fully developed and highly complex heliolithic culture.

But the drawings of pictures of objects, such as the Cro-Magnons drew on the walls of their caves, is the commencement of writing. The American Indians drew the picture of a man shooting an arrow at an animal to convey the idea of a man hunting. To show that a river was crossed the crude picture of a man crossing a river was used. This form of writing, which may be seen on rocks and cliffs east of the Sierra Nevada Mountains along the trails and at the water holes all the way from Mexico to Canada, is called pictograph writing.

In Inyo County, California, the desert heat makes it imperative that the traveler find a water hole at the end of his day's journey. The Indian routes across the Inyo Range, several of which this writer followed on foot (as these were too difficult for a horse) in 1911 and 1912, would be unnoticed by one unfamiliar with Indian methods. But from the start, if one looks ahead as far as he can see he will discern a little rock perched on a larger rock. When he reaches this place, if he looks ahead about as far as he can see he will discern another small rock perched on a larger rock. These mark the route, and by following them one is not led into a box canyon or to an impassable declivity. On the top of the Inyo Mountains big horn sheep abound even to this day. The trails on the pictograph reproduced on page 144 show where big horn sheep may be found, and the water holes (which are far apart in this region) where those hunting them can camp.

In true pictograph it is very difficult to convey abstract ideas of any kind. Its use, therefore, is quite limited. But where there is also a spoken language it is but a step, though a long step and an exceedingly important one, to unite the picture of something having the same name as an abstract idea with the abstract idea, using the picture of the object to represent the more general conception. Thus the sound of not, and knot, is identical. It is impossible to draw a picture of the general conception "not." But it is easy to draw a picture of a knot in a rope. This picture then, because the words sound the same, becomes the symbol of "not" as well as of knot. The picture of the knot is then said to be a phonetic sign.

Likewise, as James Henry Breasted points out, the picture of a "bee" may be phonetically used to stand for "be," and the picture of a "leaf" may be used to stand phonetically for the syllable "lief." The picture of a bee followed immediately by the picture of a leaf then becomes the phonetic sign for "belief." These phonetic signs so used become real writing, such as the Egyptians employed. Professor Breasted holds that such writing arose in Egypt earlier than anywhere else in the world. But the Sumerians also possessed writing as far back as we have records of them, so it is difficult to decide which developed writing first. And there is much to indicate that both derived the general idea from a common source.

However, the writing of the Sumerians, while developed along the lines above illustrated, was very different from the Egyptian system, so different that it seem certain one system was not derived from the other. The Egyptians used pictures, called hieroglyphics. They also developed at a very early date a true alphabet in which a sign signifies a single letter instead of a syllable. This true alphabet contained 24 letters, being the earliest known alphabet, and the one from which our own has descended. At a much later date there was another alphabet developed of 22 letters, each letter being an abbreviated and conventionalized symbol for one of the 12 zodiacal signs or one of the 10 planets. The religious works and other important doctrines, however, were mostly written in hieroglyphics, which was the phonetic writing employed even at a late date by the priesthood.

Another method of primitive writing, usually employed in association with hieroglyphics or pictographs, is the use of ideographs. An ideograph is the use of a picture which is not that of the object but which symbolizes the object. Thus a feather may be used as an ideograph for a bird, a crescent as an ideograph for the moon, or a hawk as an ideograph for the sun god. Ideographs were employed both by Egyptians and Sumerians, as well as by the early people of other lands, but their method of using ideographs, as well as their method of writing in general, was very different. The Egyptians cut the pictures of their hieroglyphics into stone, and wrote them in black ink, and painted them in colors upon papyrus. The Sumerians did not have papyrus, but wrote upon thin tablets of clay.

Their method of writing was to use a clay tablet while it was yet soft. They wrote with a square tipped reed, or stylus, pressing a corner of this square tip into the soft clay for each line of the picture sign. Lines produced in this manner were broad at one end and pointed at the other, tending to be wedge shaped. Each picture was thus a group of wedges. Hence the writing is called "cuneiform," meaning wedge shaped.

This cuneiform writing of the Sumerians evidently developed from pictographs to the use of hieroglyphics and ideographs. These became so conventionalized that their meaning seems quite arbitrary. They even verge closely to an alphabet. But the Sumerians never attained a true alphabet. In this way they were surpassed by the Egyptians. Yet even though the writings of the Sumerians and the Egyptians passed through parallel stages of development, in the end the characters of the writing, as well as the materials used, were so different that it is difficult to imagine that one system influenced the other.

The Egyptians made no use of clay as a writing material. Instead, they made use of a tall rush-like plant, the papyrus plant, which they cut into thin longitudinal slices and gummed together and pressed to form a sheet called papyrus. From this comes our word "paper." This papyrus, when prepared, formed a thin sheet which was usually something over twelve inches in width and from a few feet to over a hundred feet long. It was kept in the form of a scroll, and as written upon the part containing the writing was rolled up and the blank part unrolled. It constituted very convenient and durable writing material, papyri that was written on several thousand years B.C. still being in a good state of preservation.

Early Man

In whatever region the existing species of man developed, whether from Australopithecus or some other ancient strain, it seems certain that there were succeeding waves of migration from this early home at widely separated intervals. A great wave of this primitive population left the parent group not less than 700,000 years ago and gradually filtered around the world south of the equator developing ultimately those characteristics which we recognize in the Negroid races. Then sometime later, possibly about 600,000 years ago, another great wave left the parent group and gradually spread around the world along the northern coastal lines, finally developing those characteristics that we recognize in the Mongolian-Amerindian races. Still later a third wave left the parental stock and in time encircled the globe somewhat north of the equator. This people developed the characteristics that came to mark the White race.

The physical and mental differences we observe in the races of man on earth today imply no unusual pressure from the outer-plane environment or the inner-plane environment. Although Australopithecus was a pigmy only four feet tall, and we may assume black of skin, a change in environment, including his food and factors of heat and cold and astrological energies might very well, in the course of time act upon his endocrine glands sufficiently to develop the characteristics of all people existing today. While psychokinetic energies may have played their part, no mutation need to have taken place.

The thyroid gland controls the development of the skin, the amount of hair, the moisture in the skin, and the development of the bones of the skull and extremities. It is well known that giants are due to an excess of the hormone of the front pituitary gland that responds to Saturn. The pygmies, if their front pituitary hormone was stimulated would have grown to larger size. On the other hand, pygmies commonly result from over activity of the pineal gland. Thyroid deficiency leads to obesity. Cortin, the hormone of the adrenal cortex, acts upon the pigment cells of the skin, blunting their sensitiveness to light. When there is deficiency of this secretion in a white person, the skin becomes increasingly sensitive to light, bronzing until, if the deficiency is marked, there is the characteristics of Negroid skin. Furthermore, when there is a certain deficiency of the thyroid gland, it produces in a white person decidedly mongoloid features.

For the past 30 years my home has been within four city blocks of the Lincoln High school in Los Angeles. Attending this school on equal terms, and mingling freely with each other are White, Chinese, Japanese, Mexican and Negro youths. The parents of the Chinese and Japanese youths, mostly working in markets not far away, usually are small in comparison to White Americans. But their children are not small. In size, scholarship and athletic ability they compare favorably with the Whites. And most of them are very much better looking than their parents. Furthermore, White children raised in this region commonly grow to larger size than their cousins who are raised in the colder states to the east. The sunshine, the mild climate, and the accessibility to vegetables and fruits the year round from which essential vitamins are obtained, are modifying both the size and the features of children whose Oriental parents probably before coming to America lived on a quite inadequate diet.

I have already mentioned that the three chief types of the White group of humanity finally settled in Europe. Along the Mediterranean coast of Africa was another race of Dark Whites, called Berbers; still other Dark Whites were the aborigines of Egypt, while to the south of these were the Ethiopians, also Dark Whites. Crossing the Red Sea to the east of Africa into Arabia there were other Dark Whites, the Semites, typified by the Arabs and the Hebrews. Still further east in India were still other aborigines, the original Dark White population, called Dravidians. And still further to the east of these, at least at a later date, and spreading across the islands of the Pacific, were still other Dark Whites, the Polynesians. All this great group of white people possessed, when first we find records of them, religious ideas evidently derived from some common source.

They came from some locality where they had the opportunity of developing much above the culture of the old stone age. There are three chief theories as to the locality in which they developed their culture. It is believed, on geological and other evidence, that the Mediterranean Sea, before the ice cap of the last glacial period receded, was not connected with the ocean, but was merely two fresh water lakes fed by the rivers of Europe and Africa. Much of what is now the Mediterranean Sea was then a fertile valley. Some think it was in this now submerged Mediterranean Valley that the White people gained their early culture. With the melting of the vast glacial ice cap, the level of the ocean rose and cut a channel through the Straights of Gibraltar, submerging this valley and connecting it with the Atlantic.

Another theory is that the early culture of the White people was developed in Egypt. But the preponderance of the evidence goes to indicate, I believe, that culture was brought to Egypt from some other region.

The third theory is based upon the tradition of lost Atlantis. Tradition has it that when darkness settled over Atlantis and Mu colonists were sent to what later became the seven centers of ancient civilization. Deep sea divers have done some exploring on the ridge where Atlantis is supposed to have been located without finding evidence of any people having lived there. Yet there is considerable concrete evidence that both Atlantis and Mu had existence and civilization. Atlantis, by Ignatius Donelly, The History of Atlantis, by Lewis Spence, and The Problem of Lemuria, by Lewis Spence, present this evidence in detail.

But wherever it developed, the very first records of Egypt, India, Crete, Peru, Mexico, China and Mesopotamia show a high degree of knowledge, and a culture similar enough to make it certain it was developed at a time when there was close communication between those who developed it, and dissimilar enough to make it equally certain that those whose earliest records are uncovered had been isolated from each of the other six centers of civilization for a long time. Even as the same species of animal when members move to a different environment and are isolated from each other over a long period of time develop quite distinct subspecies, so isolated cultures, although retaining their main features, after a time take on quite different minor characteristics.

Extrasensory Communication

While there is similarity enough between the seven ancient centers of civilization to make it certain their cultures and their religious beliefs were developed from a common source, it is difficult to appraise the extent to which each had actual physical contact with that source. The tradition is that, as mentioned, the astrologers of Atlantis foreseeing the probable destruction of that land, were instrumental in having a colony set up in each of the regions which later became one of the seven ancient centers of civilization; even as precautionary actions are being taken in this year of 1949, that The Brotherhood of Light lessons shall survive for future generations in the event that much of the population of the globe is exterminated in a possible atomic war.

But there are many recorded instances in modern times of important new discoveries in science, important theories, and important inventions which have been made almost simultaneously by those, often in different countries, who have been quite unaware of the work being done along the same line by others. All of us who have been interested in science have read announcements of the same discovery made almost at the same time by scientists who have had no outer-plane contact with each other. Practically the same theory about some phenomenon often is published in different countries by people who do not even know of each other's existence. And the same invention is so frequently sent to the patent office by those who live long distances apart that this is one of the big problems faced by those in this government service.

There can be no doubt that extrasensory perception often plays a considerable part in causing inventions to be almost identical in detail. But that both should be working on the same problem at the same time in these cases is probable chiefly due to the general astrological weather.

In Chapter 6 (Serial Lesson 130) it was mentioned that the similar way in which marsupial mammals developed characteristics parallel with those of the various species of placental mammals from which they were isolated was probably due in about equal degree to the outer-plane environment and the inner-plane environment. Confronted with the problem of adapting to a similar kind of external condition, the creatures in both areas would struggle to find a solution. That they found a solution, and were able to adapt themselves successfully to the similar conditions is not surprising. But that the animals thus isolated from each other should solve the problem in an almost identical manner certainly was not due to chance. Nor, in many instances was the method used the only one which could have been made successfully to adapt to the condition.

And while, even as the marsupials and the placental mammals had a common ancestor, the heliolithic culture undoubtedly was the ancestor of all later civilizations, it must not be overlooked that even by peoples quite isolated from each other, developments from this culture along parallel lines probably was strongly influenced both by the extrasensory impressions each received from the other, and by the astrological weather stimulating the thoughts of each to follow a similar pattern.

The environment most favorable for inclining towards civilized pursuits is one in which there is a fertile land with an available water supply. For this reason the Valley of the Tigris and Euphrates became the seat of civilization in Mesopotamia, the Valley of the Nile became the seat of civilization in Egypt, the Valley of the Ganges the seat of civilization in India, and the Valleys of the Hwang-ho and Yangtse-kiang the seats of civilization in China.

Civilization not only requires a stable and never failing food supply that can be obtained close to the home, but it also requires some available building material out of which homes can be constructed. In the lower valley of the

Euphrates-Tigris was a clay which was suitable for drying into brick. Of this brick the earliest known inhabitants, the Sumerians, who were a dark white people, built their homes, made pottery, and upon thin sections made records and wrote communications.

Mesopotamia

It is still a matter of surmise which is the older, the civilization of Egypt or Mesopotamia. At Nippur, in Mesopotamia, an American expedition unearthed evidence of a city dated not later than 5,000 B.C. and thought by some to date as early as 6,000 B.C. This is earlier than anything of a similar nature found in Egypt. An inscription at Nippur says that the empire extended from what is now the Persian Gulf to the Mediterranean Sea. This is the first of all known empires, and its temples and its priest rulers are the oldest of which we have indisputable records.

In 1930, at Tepe Gawra excavations uncovered various occupation levels. Level six down is contemporaneous with the first Dynasty of Ur and mentions temples and astrological emblems. At Level 13 down, which dates more than 6,000 years ago, was unearthed the oldest temple now known. It is of advanced architecture, and the pottery motives and engraved seal stamps show not merely skill, but true artistic talent.

What astounded the University of Pennsylvania professors in charge of the work was that at such antiquity there were neither mud huts nor crude methods of life. As they stated it: "Those inhabitants of Level 13 were neither primitive nor normal; they were an abnormally gifted and wonderfully balanced people. And they left evidence of their achievements in more than one aspect of common life."

The oldest recorded religious beliefs are those of the Sumerians, written in cuneiform. These earliest religious records reveal that the people of the Tigris-Euphrates Valley believed in five chief deities: Sun, Moon, Storm-god, Vegetation deity, and deity of the Water. Each city had its patron deity, but although the names given to these patron deities differed, in attributes they were identical with these mentioned. At a very early date also, there was a deific trinity: Anu, the Heavens; Enlil, the Earth; and Ea, the Sea. At the same time, however, Anu was also considered the Sun-god, and Enlil the Storm-god corresponding to Jupiter. As the years rolled on the names of the gods changed. Thus the Sun-god in the north was called Ashur, the patron of the city of that name. Ninib was the Sun-god of Nippur, and Shamash was the Sun-god of another region. In Babylon, Marduk was the chief deity. He was originally the Sun-god, but gradually absorbed the qualities of all the various other deities of the surrounding territories. Later still he became identified with the planet Jupiter.

The Babylonians and Assyrians, nor the Sumerians and Akkadians before them, did not possess the scientific knowledge sometimes accredited to them. They had little ability to predict in advance the precise time when eclipses and other celestial phenomena would take place. According to A. T. Olmstead, Professor of History in the University of Illinois, who has been over the ground in person, they did have, however, great precision in observing and recording such phenomena. Speaking of fragments quite ancient he says that the fixed stars were mapped in three concentric circles belonging to Anu, Enlil and Ea, respectively. He assumes they possessed several machines for close reckoning of the stars, as star positions are given exact, and the circle is divided into 360°. A later fragmentary tablet has been found picturing a number of the zodiacal signs, each showing 30 vertical lines, corresponding to the 30 degrees of a sign. The position of a star is marked by a horizontal line over the vertical line, and on this tablet the position of the chief star in Libra is shown where it was in 800 B.C.

Other fragments are known from Assurbanipal's library, being copies of older originals collected by this literary monarch, showing the position of the equinox in Assyrian times when it was in Aries, showing it in the early Babylonian Dynasty when it was in Taurus, and one, called the Astrolabe, showing it in Gemini with the star positions as they were about the year 4,864 B.C.

All authorities agree that both the astrological positions and the important events coincident with them, as well as the astrological predictions made from such positions, were carefully recorded dating back to before the conquest by Sargon, about 2,750 B.C. This vast series of observations of the actual events which were coincident with certain astrological positions, were finally collected not later than 1,500 B.C. into what is known as the Anu-Enlil Series. This astrological handbook became the standard reference throughout Babylonia and Assyria, and as Jastrow says:

"It appears, therefore, that when an inquiry was put to an astrologer as to the meaning of a particular sign in the heavens, the Anu-Enlil Series was forthwith consulted, the sign in question hunted up, and copied verbatim, together with the interpretation or the alternative interpretations, and forwarded to the king with any needful explanation."

No less painstaking and thorough were the comparisons with actual events that transpired, or predictions made by liver divination. The forms shown by the liver that gave rise to the prediction were recorded, together later with the extent to which the prediction was fulfilled. What might be predicted from a given condition, both in astrology and by other methods of divination, was based upon what similar conditions in numerous instances in the past had presaged. These careful experimental methods were followed over vast periods of time.

The Sumerians, Akkadians, Babylonians, and Assyrians believed the universe to be peopled with both good spirits and bad spirits. They also believed in sorcery, the power of one person to injure another by what is now called psychokinesis. There is a vast literature given to incantations to secure the help of good spirits, to avert the influence of malevolent spirits, and to annul the influence of sorcerers.

They universally believed in a life after death. There was no hell, but ordinary persons continued in a conscious or semi conscious state after this life; a rather dour and inactive sort of existence in the dark, gloomy bowels of the earth. Those who won the favor of the gods went to the "Island of the Blest."

As far back as we have any record of their laws, and the complete legal code of Hammurabi, 2,200 B.C., has been recovered and translated (pictured in Chapter 5, Serial Lesson 137, Course 12-2, *Natural Alchemy: Evolution of Religion*), they indicate that the aim was toward strict justice, being quite as fair and humanitarian as those of the present day. The kings prided themselves upon being just and benefiting their people, and the legal decisions show a full conscientious weighing of all evidence. The ancient Sumerians, Akkadians, and later the Babylonians and Assyrians, so far as there is evidence to show, were no more moral and no less so than the people of today.

Egypt

History records that at the time of the early Sumerian kings in Chaldea the first dynasty of Lower Egypt was established, dated according to the Turin MS., 5,507 B.C. At this time no suitable calendar was in use and the Egyptian system of writing had not been evolved. Fifty kings were to reign before Menes, who commonly is considered the first historical ruler because, seven years after he came to the throne he established a calendar by which succeeding events could be, and were, recorded.

Menes, who thus established the First Dynasty, came to the throne 3,407 B.C. Khufu, whom the Greeks called Cheops, was the first king of the Fourth Dynasty. He ruled 2789-2767 B.C., up to within a few years of the date Sargon the Great founded his empire in Chaldea. According to our traditions, in the year 2,440 B.C. a group separated from the Theocracy of Egypt, and throughout all subsequent times, as a secret order, the name of which translated into English means, The Brotherhood of Light, has been perpetuated.

Amenhotep IV who ascended the throne in 1,375 B. C. through the death of his father, was born 1,388 B.C. At the age of 19 he contacted the same spiritual source from which now emanates The Brotherhood of Light lessons. This led him to perceive that the orthodox religion of Amen kept the minds of the people confused as to reality, steeped in fear, servile to the priests, and shut them from the knowledge which would lead to true spirituality, progress and happiness.

When the king became one of the Brethren of Light, he was determined to spread the gospel of The Religion of the Stars even as The Church of Light has determined to spread it throughout the world today. But having allied himself with the Legions of Light, he could no longer bear the name of Amen, a name which was associated with orthodox religion; for orthodoxy then, as now, kept people in spiritual darkness. He therefore renounced the name Amenhotep, and took the name Akhenaten, which means living in the Light. The details of his life and teachings, during the ten or eleven years he disseminated The Stellarian Religion before orthodoxy succeeded in bringing about his death, are set forth in the reference book, *Astrological Lore of All Ages*.

The orthodox religion of Egypt was no such simple system as that found in Mesopotamia. The Egyptian populace seems to have been cursed with the inability to forget or discard any belief once held. From time to time new gods and beliefs were adopted, but the old ones were not discarded.

Egypt in ancient times was divided into more than forty little principalities, called "nomes." Each nome, in addition to whatever gods it absorbed from without, had its own special god. This god frequently was worshipped in the form of some animal.

In so brief a space no attempt can be made to enumerate the gods of ancient Egypt. For this sees, A *Handbook of Egyptian Religion*, by Adolf Erman; and the works of Wallace Budge on The Gods of Ancient Egypt. These gods, for the most part, seem to be secondary deities; for as far back as we have records we find in Egypt the grand idea of Divine Unity. Thus Herodotus states that the Egyptians of Thebes recognized a single god, who had no beginning, and who was to have no end of days; a statement confirmed by the sacred texts. Corresponding to the mother-god, Ishtar, of the Chaldeans, was Isis of Egypt. Osiris seems to have been the father-god, the creator. And the son, the god of light, was Horus. In Chaldea, Tiamant was the god of darkness and evil, and the Chaldean mythology largely revolves around encounters in which Marduk, the god of light, vanquishes Tiamant. And in like manner Egyptian mythology largely revolves around encounters in which Horus, god of light, in the end defeats Set, the god of darkness.

Due to the continuous fluctuating names given the deities, sometimes it is Osiris, as god of light, who is pitted against Set. Osiris, again, was the sun of the lower hemisphere, Horus being the rising sun, Ra the sun at noon, and Kheper the sun when it produces and nourishes life. The sun thus became the symbol of the one supreme deity, and its journey about the earth typified the journey of the human soul. The soul, like the immortal Ra, the sun, when it descends into the tomb, is resurrected to lead a new life.

The outstanding feature of the religions of ancient Egypt is the universality of the belief that the soul lives in another region after the death of the physical body. In this after life the soul encounters certain obstacles to its progress, which may be overcome by adhering to a specific line of conduct and reciting appropriate hymns, prayers, and orations. The whole body of Egyptian religious literature has for its primary object the care of the soul after death. This literature is chiefly embodied in the Egyptian Book of the Dead.

Relative to these beliefs I will quote two paragraphs from E. A. Wallace Budge's, *The Book of the Dead*, a volume which may be consulted for further particulars:

"Osiris was the God-man through whose suffering and death the Egyptian hoped that he might rise again in a glorified Spirit-body, and to him who had conquered death and had become the king of the other world the Egyptian appealed in prayer for eternal life through his victory and power. In every funeral inscription known to us, from the Pyramid Texts down to the roughly written prayers upon coffins of the Roman period, what is done for Osiris is done for the deceased, the state and condition of Osiris are the state and condition of the deceased; in a word the deceased is identified with Osiris."

"The preservation of the body was of vital importance, because the dogma of Osiris taught that from it would spring the translucent, transparent, immaterial, refulgent and glorious envelope in which the Spirit-soul of the deceased would take up its abode with all his mental and spiritual attributes."

In this after life the good deeds done in the flesh are weighed against the evil deeds, and the soul is judged accordingly. Only the extremely wicked were supposed to be condemned, after much suffering, to final extinction.

The Egyptians also believed in, and practiced, what is now called psychokinesis, and was then known as magic.

In later times astrology was extensively practiced. At this date it is difficult to determine how early and how widely the priests were devoted to astrology. It is highly probable that only those who belonged to certain secret organizations were initiated into astrological lore. We do, however, have positive evidence of very early and very comprehensive astronomical knowledge, and in those days astronomy seems to have been studied only in behalf of astrology. Thus the Solar Calendar was introduced into Egypt in 4,241 B.C., and is the earliest dated event in history. The earliest year list, dated from some event as we date from A.D., is the Palermo Stone of Egypt, beginning 3,400 B.C. and continuing without a break for 700 years.

The workmanship of the temple of Denderah seems to be of rather a late period, but also appears to be built upon a succession of older buildings dating back to very early times. Upon the ceiling of this temple is pictured a great zodiac showing the vernal equinox in the sign Virgo, where it was about 13,000 years ago. The Great Pyramid of Gizeh was built by Cheops, requiring the full period of his reign, 2789-2767 for its construction. Embodied in its measurements are as precise and as great a knowledge of astronomy as possessed by modern astronomers, and also in its symbolism is embodied the ancient Hermetic Doctrines.

Crete

We are apt to consider the Greece of Homer's time as ancient; but recent excavations in Crete show there was an Aegean civilization extending into Crete, Cyprus, Greece, Asia Minor, Sicily and South Italy. This is not far from Egypt geographically, and it is apparent that some time in the past this culture developed from the same heliolithic ideas as that of Egypt, Mesopotamia and the other ancient centers of civilizations. And H. G. Wells holds that it is equally as old as that of Egypt, and that its inhabitants had become a people of sea commerce as far back as 4,000 B.C.

Although their writing has not yet been deciphered, it seems quite certain that Crete was united under one ruler, who was called Minos, about 2,500 B.C. From this time on until the disruption of the Empire about 1,400 B.C., there was a very high degree of civilization, including water pipes, bathrooms, pottery, textiles, ivory and metal inlaid work that has not been surpassed in quality, and female styles that include corsets and flounced dresses.

Peru

Somewhat paralleling the civilizations of Egypt in Africa, Crete in the Mediterranean, Sumeria in Mesopotamia, and that of China and India in Asia, though never rising to so great a height, and probably being of much later date, there developed in America two apparently independent centers of culture. These were the Peruvians in South America and the Maya in Mexico.

Both of these people were of the Mongolian-American group, and both show indisputable evidence of rising directly from the heliolithic culture. From picture writing the Maya developed a hieroglyphic script. But this was used chiefly in keeping records in connection with the calendar system. The Peruvians seem not to have gone farther than crude picture writing, but developed a system of keeping records by means of knots tied on variously colored and variously shaped strings.

It would seem that at least for a very long time there had been no more intercourse between Peru and Mexico than there had been at the beginning of their civilizations between Egypt and Sumeria. The potato, for instance, was one of the chief agricultural products of Peru, yet the Mexicans had never heard of it. Both people, however, worked in bronze and copper and in silver and gold. They were skilled in stone construction, made excellent pottery, did weaving, and were skilled in the use of dyes. Their architecture was of a very high order.

Among the Pre Incas knowledge of the stars and the spiritual teachings were in the custody of the Stellar Priests. Seven hundred years before the coming of the first Spaniards this people had a white granite city some distance down from the 14,000 foot Continental Divide of the Andes toward the Amazon. It consisted of 400 hewn stone houses. The temple was built of irregular, dissimilar, many joined stones of gigantic size fitted with Pre Inca nicety. One block in it is 14 feet long and 8 feet high. The city is called Manchu Picchu. Further details are given in the reference book, *Astrological Lore of All Ages*.

In America, among the Maya in Mexico and the Peruvians in South America, much the same beliefs prevailed in regard to magic, spirit communion, astrology, and the reward for righteousness, that we find among the Chinese. Each of these three peoples had great veneration for the movements of the heavenly bodies and ordered their lives largely through astrological considerations. When the first White men reached Peru they found at Curzco a great temple of the sun, in which was a huge sun of gold representing the figure of a human face surrounded by golden rays, so placed as to receive the first beams of the rising sun. The sun was the chief deity, the moon and stars were subordinate deities, and there was an order of Vestal Virgins. They were set apart at an early age for duty in the temple and to preserve the sacred fire there kept burning.

The religious ceremonies were numerous and elaborate and associated with festivals which were held at those times when the heavenly bodies were properly situated. They were for the purpose of securing rain, averting evil, producing abundant harvest, and other ends too numerous to mention. The burning of incense and offering of sacrifice to the various gods formed a part of these ceremonies, on very rare occasions human sacrifices being offered.

Mexico

More is known of the religion of the early inhabitants of Mexico and Central America, because the string records of the Peruvians are not so well understood as the hieroglyphics of the Maya. With the latter the calendar is by far the most important religious feature. Not only were all religious observances according to the calendar, but an elaborate system of divination was worked out by the priests, whose duty it was to note the movements of the planets. They had also observed that Venus disappears for eight days between the last glimpse of her in the West and the first glimpse of her as a morning star in the East, and that the same phases of Venus come back in about 548 days, so that five Venus years are equal to practically eight ordinary years. From this, in addition to an annual calendar, they had computed a Venus calendar.

Dr. Spinden, of Harvard, announced in December, 1925, that he had deciphered this Venus calendar, and that it was finally put in working order between two risings of Venus as morning star in conjunction with the summer solstices of 538 and 530 B.C. He has also proved that the first date on which the Maya gave each day its consecutive number, so that the records subsequently are complete, was August 6, 613 B.C., and that the perfected annual calendar was inaugurated on the winter solstice 580 B.C. Back of these perfected calendars must lie age long astronomical observations.

The sun was the chief deity of the Maya. He ruled the East. The god of storm ruled the North, the god of maize ruled the West, and the death-god ruled the South. There was a bat-god of the under world, and each day was ruled over by a special spirit. The unseen universe was peopled with such spirits, to whom sacrifices were made for special purposes. And to facilitate matters, images were made of stone, pottery, and other substances, of these spiritual entities, and the sacrifices made before the image. Such images, of course, are mere symbols of unseen presences and powers; therefore let no one who has ever knelt before a cross, or bowed his head in the presence of the image of a saint scoff at such modes of worship. There was also a system of confession by which through gifts and offering of some of his own blood obtained by pricking his veins, the penitent might obtain, through priestly intercession, absolution for sins. Sacrifices were made at time of planting, for harvest, for rain, and for many other boons. In time of national crisis, prisoners of war were offered as human sacrifices. As time passed human sacrifices became more numerous. Later, among the Aztecs, thousands of human beings were sacrificed to appease the angry gods, their hearts being torn out of the living bodies of the victims and held quivering aloft. Thus in the New World a priesthood developed which drenched the country with the blood of the innocent, even as in the Old World countless innocents were sacrificed by no less cruel means for heresy, all in the name of religion, but of religion gone to seed. Further details of the astrology of the Maya and the astrology of the Aztecs is given in the reference book, *Astrological Lore of All Ages*.

China

Our first view of China is as a great empire, or group of principalities, about 2,700 B.C. to 2,400 B.C., ruled over successively by five emperors. The people are of the Mongolian-American group. They spread from the ten fertile valleys of the Tarim down into the valley of the Hwang-ho and later to the valley of the Yangtse-kiang. They also spread from the south, where they mingled with the heliolithic culture people of Siam and Burma. Furthermore, as far back as we are able to trace them, they were in possession of writing.

This writing, from its form, evidently developed from pictographs through ideographs and phonograms. But when it reached the phonetic stage, due to the fact that the Chinese language consists of a very few simple monosyllabic sounds, it necessarily departed quite markedly from the writing developed in other lands. Thus in Chinese, there being no grammar, the same word when spoken is used to denote a large number of things. The particular picture of some object signified by the spoken word—usually a picture easily drawn—was used to signify any one of the numerous things denoted by the spoken word. In addition to this picture, representing the sound of the word, there was also necessary another picture to signify which of the various ideas was meant. It was thus possible to express practically any idea.

The second sign which gives precision to the first picture is called a determinative. Determinatives were also used in the Egyptian hieroglyphics. But the Egyptian language did not need such a complicated system of signs to express itself phonetically.

Because of the peculiar makeup of the Chinese spoken language it became necessary to use an almost unlimited number of written characters to express it. And for convenience in writing them with a brush, the pictures were abbreviated and conventionalized and represented by groups of marks each of which it was possible to make by a stroke of the brush. Such groups of brush marks gradually lost semblance to the original pictures, and it became necessary for anyone desiring to read or write Chinese to memorize a vast number of complex signs. This, because it required so much time and effort, developed a special literary class, the mandarins, who also became the rulers. And even among these it took so long to learn to read and write that it would seem they had little time left for other lines of thought. Furthermore, in learning the Chinese characters they spent so much time with the traditional literature of the country that these ideas gained a firm hold upon their minds. To such an extent is this true that it is with the utmost difficulty that they are replaced by more up to date notions.

That which is of more interest to us here, however, is that at the very commencement of authentic knowledge about the Chinese we find them with a knowledge of medicine, with a knowledge of astronomy, and in possession of the civilized arts and industries. Neither their writing nor their culture was obtained from, nor in any appreciable degree influenced by, Egypt or Mesopotamia. It seems to have been derived with them from a common source, but so far in the past that its subsequent development made of it a great contrast in every way to that developed in any other part of the world.

Study Questions

These questions do not represent the final exam for this course.
The following questions are for study purposes only.

Please Note: There is only one final examination sheet for the two sections of Course 12

Chapter 1, Origin of the Earth

1. Why is religion so important to any individual?
2. Upon what depends a perfect religion?
3. To be adequate, upon what must religion be based?
4. To what extent is man influenced by his inner-plane environment?
5. Indicate how the iron curtain of orthodoxy retarded the acquisition of knowledge of the physical world.
6. Explain how the spontaneous disintegration of radioactive minerals enables the age of rocks containing them to be determined.
7. About how old are the oldest rocks on earth?
8. What is the present most plausible theory as to the formation of the solar system?
9. Under the Carnot-Clausius law, in what direction is all inorganic evolution moving?
10. Is it consistent with probability calculations to believe that the first living cell was due to a chance combination of inorganic molecules?
11. What have psychical researchers found whenever physical conditions are present that permit the manifestation of intelligence?
12. Is there any evidence of the existence of a God who has human frailties?
13. Is there much indication that there is an all-pervading Super Intelligence?
14. How is the psychokinetic power of the mind demonstrated in university laboratories?
15. How large is the Milky Way, and how many stars does it contain?
16. How many extra-galactic nebulae are within the range of the 100-inch telescope?
17. Of what are the extra-galactic nebulae composed?
18. What type of stars are the largest, and how big is Antares?
19. What type of star is our sun?
20. What is a White Dwarf?
21. Explain the behavior of variable stars.

22. Of what are galactic nebulae chiefly composed?
23. Of what are comets composed?
24. Is the tail of a comet dangerous to the inhabitants of earth?
25. What is the cause of the slowing down of the earth's rotation?

Chapter 2, Origin & Development of Plants

1. Is it believed that the earth has a molten interior?
2. What gave rise to the great mountain chains such as the Rocky Mountains and the Andes?
3. What kind of an atmosphere had the earth at the commencement of geologic time?
4. What is the effect upon the climate of elevating land areas?
5. In what direction, following the second law of thermodynamics, does inorganic evolution move?
6. In what direction does the evolution of life move?
7. What calculations show the inconsistency of believing the appearance of the first living cell was due to a chance combination of inorganic molecules?
8. What have psychical researchers found whenever physical conditions are present that would permit the manifestation of intelligence?
9. What makes it probable that it was due to psychokinesis that the necessary molecules were combined to permit intelligence to manifest through a primitive single-celled organism?
10. Through what power does a soul attach itself to and manifest through a physical form?
11. Upon what are plants dependent for the assimilation of carbon dioxide from the air?
12. With what substance is all life on earth associated?
13. What type of life was probably the first to get a foothold on earth?
14. What are the three hereditary drives of all life?
15. In what way are animals dependent upon plants?
16. What primitive organism has the characteristics of both plants and animals?
17. Through what process does growth in both animals and plants take place?
18. What advantage was gained by the close cooperation of a colony of cells?
19. What is the anchoring device by which kelp keeps from being washed out to sea?

20. What was the condition which developed the first roots?
21. What was the desire which caused plants to build stems?
22. By what means do ferns reproduce?
23. What advantage have seed plants over those which produce no seeds?
24. Of what are the conifers supposed to be modifications?
25. In what way do fire-type pines take effective precautionary actions?

Chapter 3, Progress of Invertebrate Life

1. What must life-forms do when the environment radically changes?
2. What evidence is there that inner-plane weather affects animals in the same manner it affects men?
3. To survive is it enough that a life-form adapt itself to the outer-plane weather?
4. What is the basic form of all organs of higher plants?
5. What evidence is there that plants have memory?
6. What was the urge that brought psychokinetic power into play to provide animals with locomotion?
7. Does subjective intelligence make mistakes?
8. What were the first animals on earth?
9. How does the protozoan obtain its food?
10. What is chitin?
11. What significance has it that early plants and animals were able to secrete substances that were not protoplasm?
12. How do protozoa reproduce?
13. What advance over more primitive colonial animals was made by the volvox?
14. What advantage derives from the exchange of chromosomes?
15. How do the sponges obtain their food?
16. What advance over the sponges has been made by the jellyfishes?
17. What animals were the first to have a right and left side and a front end?

18. What advantage have the roundworms over the flat worms?
19. What advance was made by the wheel-worms over the flatworms and the roundworms?
20. What advance have the Bryozoa made over the wheel-worms?
21. In what way are the starfishes superior to the Bryozoa?
22. What in the common earthworm foreshadows a brain?
23. How does the digger wasp provide fresh meat for her young?
24. What kind of a heart do mollusks have?
25. What is the main advance over other forms made by the Chordata, which embraces the vertebrate animals?

Chapter 4, Fishes and Amphibians

1. What significance relative to progress has the inner-plane weather at the time a life-form is born?
2. What insurance is given by doubling the reproductive genes?
3. What is a mutation?
4. What effect on mutations have radioactive materials?
5. Can acquired characteristics be inherited?
6. Does the inner-plane weather markedly affect endocrine secretions?
7. Do endocrine secretions markedly influence the physical form?
8. Have the changes in form and function made by animals in their efforts to adapt themselves to a new environment always been beneficial?
9. How do insects use plants and plants use insects?
10. Upon what do all those plants depend for pollination that have small greenish inconspicuous flowers?
11. Do bees seeking nectar indiscriminately visit different kinds of flowers?
12. What insects makes biscuits and bakes them in the sun?
13. What were the first truly vertebrate animals?
14. What kind of supporting tissue would prevent the shortening of the form by the water of a stream flowing against the head?
15. What is the most outstanding characteristic of truly vertebrate animals?

16. What is the second characteristic of all vertebrate animals?
17. Why is it believed that such fish today as the shad, sturgeon and salmon leave the sea and ascend rivers to spawn?
18. How does the seahorse carry its eggs?
19. What makes a nest of leaves and stems in which the female lays her eggs?
20. What peculiar habit has the "climbing perch" which abounds in fresh water throughout the Malaya countries?
21. Psychokinesis would try to bring about what change in form to adapt to the environment in which African mud fish live?
22. In what geologic period are the first fossil fish found?
23. What is the oldest fossil amphibian?
24. In what environment must amphibians lay their eggs?
25. What environmental condition caused psychokinesis to work to develop amphibians from fish?

Chapter 5, Reptiles and Birds

1. Upon what principle is natural selection based?
2. At the present time how much land is required to raise enough food for one person?
3. How many acres of land are there in the world per person that can be used for food production?
4. Illustrate competition for food between different species of animals.
5. Illustrate competition for food between individuals of a single species.
6. What is a third form of competition?
7. Does the gradual development of some organ which is advantageous when perfected always give an advantage in its earlier stages of development?
8. What does changing inner-plane weather insure?
9. Toward what type of form, generally speaking, has all life on earth been evolving?
10. What are the two periods following the Devonian considered together called?
11. What kind of conditions in the Lower Carboniferous Period favored the formation of coal?
12. What were the conditions and the life of the Upper Carboniferous period?
13. What climatic change took place at the end of the Upper Carboniferous period?

14. What was the advantage of an egg having a more ample yolk?
15. What was the advantage of an egg having a shell?
16. How did the insects adapt themselves to the cold of the ice age of the early Permian times?
17. How does the Yucca Moth insure food for her young and for her future generations?
18. What kind of climate was prevalent during the Age of Reptiles?
19. How did aridity affect reptile habits?
20. Why is it more difficult for fish and reptiles than for mammals to adapt themselves to heat and cold?
21. How did creatures, when the cold shut down, solve the problem of keeping warm?
22. What habit of certain dinosaurs may have assisted in developing the power of flight?
23. An adaptation of what earlier covering are the feathers of birds?
24. Were birds the first creatures to solve the problem of flight?
25. What were the two divergent races of dinosaurs that developed before they became extinct?

Chapter 6, Development Among Mammals

1. What kind of mammals existed before the Cenozoic era?
2. How long before the Cenozoic era had mammals remained sly little creatures always on the dodge?
3. Are most mammals of today of the marsupial or the placental type?
4. Name a mammal that lays an egg?
5. How is the milk of the duck mole obtained by it's young?
6. What is the most familiar marsupial mammal of today?
7. Of what advantage is the placenta in mammals of the type possessing it?
8. Why is it believed that in the Cretaceous period all of the world were connected by land?
9. How large were the mammals of 50 million years ago?
10. Were there members of the dog or cat families in Basal Eocene times?
11. Where did the horse have its origin and most of its development?
12. How large was Eohippus?

13. How many toes on each foot had Eohippus?

14. From what kind of creatures of the Cretaceous period did the Primate mammals develop?

15. Why during the Oligocene period could mammals migrate both ways between America and Europe?

16. In what period did the dog group separate from the cat group?

17. What was the period of greatest mammalian abundance?

18. Why did the aridity of the Miocene period affect the teeth of horses then living?

19. About when did the Orangutan branch off from the common primate stem?

20. About when did the Chimpanzee and the Gorilla branch from the common primate stem?

21. Since it branched off has the Chimpanzee become more, or less, human?

22. Since it branched off has the Gorilla become more, or less, human?

23. Since the Middle Miocene have the living apes developed along lines similar to that of man?

24. During what period is the first of modern horses found?

25. About how many years ago did the Age of Ice of the Pleistocene period develop, and how thick was the ultimate ice sheet that covered what is now New York?

Chapter 7, Development of Man

1. To what group of mammals, does man belong?

2. In formation, about how old are the first Primate fossils found?

3. For about how long has there been a constant widening of the structural breach between man and the living apes?

4. What is the earliest man whose fossils have as yet been discovered?

5. What is supposed to have been the environmental condition which influenced this earliest man to develop the erect posture?

6. What is supposed to have been the environmental condition that is one of the reasons the gorilla has remained in development where he was 15 million years ago?

7. Is there any evidence man lived on earth prior to the Pliocene period which commenced about seven million years ago?

8. Cite proof that animals have extrasensory perception.

9. Within man's anatomy are about how many vestigial structures?

10. What do those vestigial structures indicate?
11. What is indicated by the human embryo being first like an invertebrate, then like a fish, then partly reptile-like and partly bird-like, then like a mammal, and finally like a man?
12. How does blood transfusion afford proof as to man's nearest kin among the animals?
13. What is indicated as to man's nearest kin by the precipitin test?
14. About how long ago did the Ape-Man of Java live?
15. About how long ago did the Peking Man live?
16. When and where did the Heidelberg Man live?
17. When and where did the Piltdown Man live?
18. Up to how long ago did the Neanderthal Man live?
19. Which of these men is called the Cave Man?
20. Does modern man belong to the same species as any of the men thus far mentioned?
21. How many species of men are there on earth today?
22. What are the three subspecies of man living on earth today?
23. What was the first of the modern type of man to appear, and when did he appear?
24. What type of people migrated into Europe about 12,000 years ago?
25. Did the Nordics come later or earlier, and where did they come from?

Chapter 8, Development of Knowledge

1. Did the Cro-Magnon possess any form of writing?
2. What kind of writing did the early American Indians use?
3. Illustrate the use of phonetic signs.
4. Was the system of writing used by Early Egyptians and Early Sumerians similar?
5. What is an ideagraph?
6. On what did the Egyptians write with black ink and in color?
7. How did the Sumerians proceed in cuneiform writing?
8. About how long ago did the Negroid races filter around the world south of the equator?

9. About how long ago did the Mongolian-Amerindian races spread around the world along the northern coastlines?

10. Illustrate how certain glands influence size, structure and complexion.

11. Illustrate how food and climate influences the size and appearance of people.

12. What kind of culture did the white people of the world possess when we find first records of them?

13. Name the seven ancient centers of civilization.

14. How does extrasensory perception probably often play a part in simultaneous invention?

15. How far back were careful records made of astrological positions, the predictions made from them, and the events that actually happened coincident with them?

16. At about what date were those observations collected into the astrological handbook known as the Anu-Enlil Series?

17. How does the legal code of Hammurabi, of about 2,200 B.C., compare with that of the present day?

18. About when was the first dynasty of Lower Egypt established?

19. What did Akhenaten attempt to do?

20. What is the outstanding feature of the religious of ancient Egypt?

21. How much civilization had ancient Crete?

22. What was the chief deity of the ancient Peruvians?

23. About when was the Venus calendar of the Maya put in working order?

24. When was the perfected annual calendar of the Maya inaugurated?

25. What kind of a nation was China about 2,700 B.C.?

Please Note: There is only one final examination sheet for the two sections of Course 12